RECENT ADVANCES IN ENGINEERING

First edition published 2024
by CRC Press
4 Park Square, Milton Park, Abingdon, Oxon, OX14 4RN

and by CRC Press
2385 NW Executive Center Drive, Suite 320, Boca Raton FL 33431

CRC Press is an imprint of Informa UK Limited

British Library Cataloguing-in-Publication Data
A catalogue record for this book is available from the British Library

ISBN: 9781032656847 (pbk)
ISBN: 9781032657271 (ebk)

DOI: 10.1201/9781032657271

Typeset in Sabon LT Std
by Ozone Publishing Services

PROCEEDINGS OF THE INTERNATIONAL CONFERENCE ON RECENT ADVANCES IN CIVIL ENGINEERING (ICRACE) 2022, DECEMBER 01-03, 2022, KOCHI

RECENT ADVANCES IN CIVIL ENGINEERING

Edited by

Dr Job Thomas & Dr Subha Vishnudas

 CRC Press
Taylor & Francis Group
Boca Raton   London   New York

CRC Press is an imprint of the
Taylor & Francis Group, an **informa** business

Contents

Theme: Building Technology

Theme: Environmental Engineering

Contents

Theme: Geotechnical Engineering

Theme: Water Resources and Hydraulics Engineering

Theme: Structural Engineering

List of Figures

List of Tables

Editors

Prof (Dr) JOB THOMAS
Professor in Civil Engineering,
Cochin University of Science & Technology,
Cochin 682022, Kerala, India
Email: job_thomas@cusat.ac.in

Dr Job Thomas is a renowned consulting structural engineer and is currently working as Professor and Head of the Department of Civil engineering at Cochin University of Science and Technology. He has published more than 50 research papers in national and international journals. He received the outstanding Concrete Researcher Award in 2020 from the Indian Concrete Institute, Kochi Centre.

Prof (Dr) SUBHA VISHNUDAS
Professor in Civil Engineering
Cochin University of Science and Technology,
Cochin-682022, Kerala, India
Email: v.subha@cusat.ac.in, subhakamal@gmail.com

Dr Subha Vishnudas, Professor in Civil Engineering, Cochin University of Science and Technology, Cochin, Kerala, India. She graduated in Civil Engineering in the year 1991 and took Master's in 1995 from Kerala, India. She pursued PhD from Delft University of Technology, The Netherlands in 2006. She has about 25 years of teaching and research experience and has around 80 publications in international journals and conferences. She has visited several countries like The USA, Germany, Australia, Japan, Netherlands, Spain, etc. for papers presentation and invited talks. She is a member in several professional bodies in India and abroad. About 10 PhD research scholars are working under her guidance in the field of sustainable development. Her area of interest include water resources, sustainable building materials, sustainable water management, sustainable waste management, housing and green buildings.

Foreword

The growth in civil engineering is very fast and there are many opportunities and challenges in this area. It is important that the opportunities are to be transformed into beneficial practices among engineers, material managers and other officers involved in the planning, design and execution of the infrastructure projects. The growth of any country largely depends on the development of infrastructure within and hence, the advancements in civil engineering field contributes towards the GDP of the country.

The Civil Engineering department of Cochin University of Science and Technology organized an International Conference on Recent Advances in Civil Engineering (ICRACE) to disseminate the know-how and challenges in this area among technocrats, practicing civil engineers, researchers etc. This conference has been conducted biennially since 2004. The conference holds an interactive platform to find solution for various problems in construction field.

This book is the collection of selected high-quality papers presented in International Conference on Recent Advances in Civil Engineering, Kerala, India, December 2022. These papers were selected based on rigorous review process by respective expert panel in theme areas. The various themes of this conference include building technology, environmental engineering, geotechnical engineering, structural engineering and water resource engineering. The invaluable contributions of all those involved in bringing this book to light is acknowledged.

All of chapters presented in this book have been prepared directly from the camera ready copy of the manuscript submitted by the authors and editing has been restricted to minor changes only where it was considered absolutely necessary

Kochi-22

Job Thomas & Subha Vishnudas
Editors

Theme: Building Technology

1. A Critical Review on the Sustainability Compatibility of Building Rating Systems

Divya Mohanan* and Deepa G Nair
Cochin University of Science & Technology, India
*Corresponding author: divyadeepakk@gmail.com

ABSTRACT: In the last decade, the advancements in the built environment have led to the evolution of many building rating systems that address sustainability of the building in one or the other way. The most recognized definition of sustainable development from the Brundtland Report states that development meets the needs of the present without compromising the ability of future generations to meet their own needs. This definition serves as a foundation for many fields including the building sector to consider sustainability and focuses on the many pillars of sustainability such as social, economic, technical, cultural, and environmental. The building industry due to its multifaceted nature requires building codes, standards, and certification systems to effectively address the sustainability assessment. This chapter attempts to put forward an extensive literature review of seven popular building rating systems: LEED (US), BREEAM (UK), CASBEE (Japan), GRIHA, LEED, IGBC, scrutinizing their macro areas, segments of sustainability and thus highlight the need for a framework which addresses the assessment of the building in terms of sustainability as a whole.

KEYWORDS: Building Rating Systems, Sustainability, LEED, BREEAM, CASBEE, GRIHA, LEED, IGBC

1. INTRODUCTION

The building industry plays a fundamental role in structuring the usage of resource consumption from extraction of raw materials to the construction and disposal phase and hence, is responsible for huge solid waste generation, environmental damage, and approximately stand third in global greenhouse gas emissions (Li et al., 2017). Policies are needed to control the negative impacts of construction on the environment. According to the UN, 1992, addressing environmental issues alone is however insufficient because the construction industry also has the responsibility to ensure economic and social developments [2]. The building plays an important role in addressing basic human needs through the provision of housing and social infrastructure facilities. The introduction of the concept of sustainability in construction surpasses environmental sustainability (Green Agenda) to incorporate economic and social sustainability (Brown Agenda), which emphasizes possible value addition to the quality of life of individuals and communities, that is, decent and comfortable housing is necessary for all citizens (du Plessis, 2007). The social perception of housing in the Brown Agenda is to provide shelter, sense of safety, and security to a community. The economic perception is that the construction of a house significantly contributes to the economy of the building industry thereby increasing the GDP yearly and the environmental perspective is to reduce the impact of the housing on the environment (Syed Jamaludin et al., 2018). Thus, the housing industry can play a significant role in promoting sustainable development and due to its multifaceted nature, building codes, standards, and certification systems are necessary to effectively address sustainability.

Over the past decade, in response to the growing awareness of sustainable construction, there has been a plethora of building performance assessment systems emerging to measure how well or

Chapter 1 DOI- 10.1201/9781032657271-1

poorly a building is performing, or is likely to perform, against a declared set of sustainability criteria (Li et al., 2017; Tang et al., 2020; Ziabakhsh and Bolhari, 2012). The developed countries' assessment emphasis is to maintain standards of living while reducing resource depletion and environmental damage (Okeyinka, 2014). The developing countries including India recognized the importance of building assessment systems for improving the sustainability of the building sector. Consequently, they adopted or customized some building performance assessment systems from developed countries (Vlasin, 1973). However, in the case of developing countries, the average standard of living is far lower than in developed countries and in many cases, basic human needs are not being met (Gibberd, 2002; Mateus et al., 2005; Schultmann et al., 2009). The emphasis in developed countries could therefore be on development that aims to address and improve the sociocultural and economic aspects along with avoiding negative environmental impacts. ISO 15392:2019 (Sustainability in Buildings and Civil Engineering Works) identifies and establishes that the sustainable construction works should consider sustainable development with respect to the three primary factors (social, economic, and environmental), along with meeting the requirements for technical and functional performance.

This chapter aims to investigate the compatibility of the categories of popular rating systems with the indicators of sustainability. Thus, highlight the need to enhance the sustainability level of the rating systems such that they evaluate all the aspects of building sustainability, especially in the developing countries.

2. BUILDING RATING SYSTEMS

In the early 1990s, building environmental assessment methods came into existence. The first building rating system was BREEAM (Building Research Establishment Environmental Assessment Method) developed in the UK in 1990. Since then, countries all over the world started developing rating systems to evaluate and certify the sustainability of buildings. Selection criteria for the rating systems addressed in this chapter are based on the number of countries implemented, the number of certified projects, and having more than 10 years of status. LEED, BREEAM, and DGNB adhered to all three criteria. CASBEE was selected on the basis of the number of certifications and more than 10 years of development. Three of India's primary rating systems, GRIHA, LEED-INDIA, and IGBC, were also selected and a review of the same is given in the next section.

2.1. Review of the selected rating systems

This section discusses the credits or weightages achievable on the basis of specified criteria and the building certification attained according to the reached total score.

2.1.1. BREEAM

BREEAM (Building Research Establishment's Environmental Assessment Method) uses a benchmarking scheme for the level of certification award. BREEAM has 145 total points, obtained including only the residential building items. Energy is the most credited category with a maximum of 31 points followed by the human well-being category which is awarded 22 points. The resulting overall score is then translated into a rating on a scale of BREEAM® certification levels: pass, good, very good, excellent, and outstanding.

2.1.2. LEED-United States and India

LEED (Leadership in Energy and Environmental Design) is the most widely used green building rating system in the world. Available for virtually all building types, LEED addresses energy efficiency, water conservation, site selection, material selection, daylighting, and waste reduction. The highest awarded category is "Energy and Atmosphere" with 37 points followed by indoor environmental quality with 18 points. LEED has four levels of certification, depending on the score reached by a building. LEED

(US and India) provides a maximum achievable total score of 100 with an extra 10 bonus points in the innovation and regional priority category. These levels are certified (40–49 points), silver (50–59 points), gold (60–69 points), and platinum (80+ points).

2.1.3. CASBEE

CASBEE (Comprehensive Assessment System for Built Environment Efficiency) was developed in Japan in 2001. CASBEE presents a new concept for assessment that distinguishes environmental load from the quality of building performance. It is certified according to the Bureau of Energy Efficiency (BEE) indicator which is the ratio of "Building Environmental Quality and Performance" (Q) and the "Building Environmental Load" (L). Q and L are the two assessment categories which are in turn divided into subsystems Q1, Q2, and Q3 and LR1, LR2, and LR3, respectively, with different weights for assessment. Each criterion is scored from levels 1 to 5, with level 1 defined as meeting minimum requirements, level 3 defined as meeting typical technical and social levels at the time of the assessment, and level 5 representing a high level of achievement. CASBEE has a total of 10 points.

2.1.4. IGBC

The Indian Green Building Council (IGBC), part of the Confederation of Indian Industry (CII), was formed in the year 2001. The rating system evaluates certain mandatory requirements and credit points using a prescriptive approach and others on a performance-based approach. The energy category is the highest credited category with 28 points followed by water conservation with 18 points. This system awards a rating of buildings as certified (40–49 points: good practices), silver (50–59 points: best practices), gold (60–75 points: outstanding performance), platinum (75–89 points: national excellence), and super platinum (90–100 points: global leadership).

2.1.5. GRIHA

Green Rating for Integrated Habitat Assessment (GRIHA) was developed by TERI (The Energy and Resources Institute) and supported by the Indian government's MNRE (Ministry of New and Renewable Energy). GRIHA adopts the five "R" (Refuse, Reduce, Reuse, Recycle, and Reinvent) philosophy of sustainable development. It is a 100-point system consisting of some mandatory core categories. One star is for 25–40 points, two stars (41–55 points), three stars for (56–70 points), four stars (71–85 points), and five stars (above 86 points).

2.1.6. DGNB

DGNB (Deutsche Gesellschaft für Nachhaltiges Bauen) abbreviated in German for the German Sustainable Building Council. DGNB is a nonprofit organization based in Stuttgart founded in 2007. Assessments revolve around the three core sustainability factors: environmental quality, economic quality, sociocultural and functional quality, and technical quality, each with an equal weightage of 22.5. An additional 10 credits were given for process quality giving a total sum of 100. A DGNB certificate in platinum (65–80%), gold (50–65%), silver (50–35%), or bronze (35% and less).

2.2. Sustainability analysis of selected rating systems

The triple bottom line concept of sustainability, coined in 1994 by John Elkington, addresses the importance of balance between the environmental (ENVS), social (SCS), and economic factors (ES) of sustainability, also known informally as 3Ps (profits, planet, and people) [12]. The categories which do not have a direct linkage to social, economic, and environment are included under the other aspects. As all the rating systems except DGNB are environmental performance assessment systems, most of their categories have compliance with environmental sustainability. The categories which have compliance with the social and economic aspects are placed accordingly. Figure 1 helps to visually identify the average weightage compliance of each rating system with 3Ps.

2.3. Environmental sustainability

From Table 1 and Figure 1, it is seen that the planet aspect of sustainability is given high priority in all the rating systems. Energy, indoor environmental quality, sustainable sites, water efficiency, pollution, and building materials are the main categories assessed by all the rating systems. The energy criterion is associated with energy use, efficiency, monitoring, and renewable energy utilization. The criterion sustainable site focuses on the influence of the site location and characteristics on the building; water category considers aspects related to the total effective water use; material category assesses the use and impacts of building materials from cradle to grave; indoor environmental quality assesses the issues that mainly influence the indoor human well-being and functional characteristics of interior spaces [13, 14]. Energy category is given the highest credit in the rating systems followed by building materials and indoor environment quality. Water is given higher credit in IGBC and GRIHA compared to LEED and BREEAM. The impact of materials from cradle to grave, which can be assessed by means of life cycle analysis is evaluated only in BREEAM, DGNB, and GRIHA. This aspect can be included in the other rating systems considering that allow the comparison of products on the basis of the same functional quality. Only CASBEE takes into account the global warming, heat island effect, light and noise pollution, and NO_x and CO_2 emissions in indoor environmental quality. Also, the operational waste management is considered only by BREEAM, CASBEE, and GRIHA. Hence, it is arguable that these tools are comprehensive enough to evaluate the environmental sustainability but can be made more effective.

2.4. Economic sustainability

The economic aspect of sustainability is considered only in DGNB and GRIHA in terms of life cycle cost in the former and monetary values in the latter. None of the rating systems address the economic or profit aspect of sustainability. Ali et al., Wei et al., and Cidel observed that the affordability aspect of the users and life cycle costing are important factors affecting economic sustainability but are given no consideration in the rating systems.

2.5. Social sustainability

The people aspect of sustainability, that is, the social sustainability factors such as safety, user satisfaction, and security are taken into account in DGNB but not prioritized in other rating systems. Though BREEAM and LEED consider some social aspects like walkable streets, transit facilities, and amenities proximity, it is not mandatory (Chiu, 2002; Vlasin, 1973).

2.6. Other aspects

The factors which enhance the design robustness such as durability, serviceability, and functionality are considered to be given equal credit in DGNB under the criteria of technical quality. Almost all rating systems have innovation category which provides points for innovations in design but does not specify the sub-categorization. As these factors can be directly related to technology, it may be considered as a fourth aspect.

From Figure 1, it is seen that energy efficiency is the main criterion for all the rating systems followed by indoor environment quality, sustainable sites, water, and resource efficiency. The environment pillar with an average of 65% is the dominant aspect in the rating systems. The average credit for social and economic pillars is very low at approximately 9%. LEED and GRIHA grand an extra 10 and 5 credits, respectively, for innovation. Only DGNB deliberates social, economic, and technological aspects of sustainability. In LEED, CASBEE, IGBC, GRIHA, and to an extent CASBEE, significant weightage is given to a single criterion, and thus there is a possibility of obtaining a green certification by only fulfilling one particular criterion even though all the other key credit criteria may be overlooked or completely ignored. The LCA aspects, operational waste, global warming heat island effect, noise, and light pollution which are now considered only by some rating tools can be implemented in all the other rating systems to improve environmental sustainability.

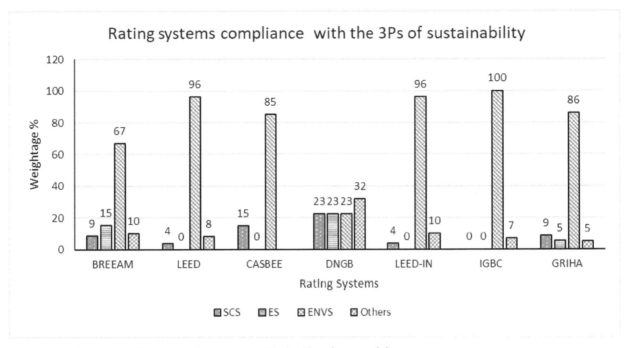

Figure 1. Weightage compliance of rating systems with the 3Ps of sustainability.

3. CONCLUSIONS

Buildings need to be perceived in line with sustainability principles as it plays an important role in improving sustainable development. This research has identified many green building rating tools used worldwide. However, due to the large extent of its usage, the analysis was limited to seven building rating systems. These rating systems were analyzed using their macro areas with the indicators of sustainability and it is seen that environmental sustainability is given the maximum priority. Social and economic indicators are currently given very low priority in the rating systems. It can thus be seen that the existing rating systems are single-dimensional in their framework and inadequate in addressing the concept of sustainability as a whole. Developing countries were found to have no significant differences from those from developed countries in terms of their scope of assessment. According to the definition of sustainable development (UN, 1992), there should be a balanced and holistic approach to the various pillars of sustainability. Thus, there is a need for the development of a framework which can be holistic, more comprehensive, and keen in terms of the range of issues addressed.

REFERENCES

[1] Y. Li, X. Chen, X. Wang, Y. Xu, and P. H. Chen, "A review of studies on green building assessment methods by comparative analysis," *Energy Build*, vol. 146, pp. 152–159, 2017, doi: 10.1016/j.enbuild.2017.04.076.

[2] C. du Plessis, "A strategic framework for sustainable construction in developing countries," *Constr Manag Econ*, vol. 25, no. 1, pp. 67–76, 2007, doi: 10.1080/01446190600601313.

[3] S. Z. H. Syed Jamaludin, S. A. Mahayuddin, and S. H. A. Hamid, "Challenges of integrating affordable and sustainable housing in Malaysia," *IOP Conf Ser Earth Environ Sci*, vol. 140, no. 1, 2018, doi: 10.1088/1755-1315/140/1/012001.

[4] D. Tang, K. Ho, and H. Foo, "A review of the green building rating systems," no. November, 2020, doi: 10.1088/1757-899X/943/1/012060.

[5] N. Ziabakhsh and M. Bolhari, "Sustainable rating systems in buildings : An overview and gap analysis," vol. 4, no. 3, pp. 226–228, 2012.

[6] Y. R. Okeyinka, "Housing in the third world cities and sustainable urban developments," *Dev Ctry Stud*, vol. 4, no. 8, pp. 112–121, 2014, http://iiste.org/Journals/index.php/DCS/article/view/12204

[7] R. D. Vlasin, "United Nations Conference on human environment," *J Community Dev Soc*, vol. 4, no. 1, pp. 22–28, 1973, doi: 10.1080/00103829.1973.10877486.

[8] C. R. Mateus et al., "Use of rating systems in the process towards sustainable construction," pp. 51–97, 2005.

[9] F. Schultmann, N. Sunke, and P. Kruger, "Global performance assessment of buildings: A critical discussion of its meaningfulness," *SASBE2009 - 3rd CIB International Conference on Smart and Sustainable Built Environments*, 2009, http://www.sasbe2009.com/proceedings/documents/SASBE2009_paper_GLOBAL_PERFORMANCE_ASSESSMENT_OF_BUILDINGS_-_A_CRITICAL_DISCUSSION_OF_ITS_MEANINGFULNESS.pdf

[10] J. Gibberd, "The sustainable building assessment tool assessing how buildings can support sustainability in developing countries," no. May, 2002.

[11] L. Bragança, R. Mateus, and H. Koukkari, "Building sustainability assessment," no. July, 2010, doi: 10.3390/su2072010.

[12] M. A. Fauzi, "A comparison study on green building tools," *International Conference on Architecture and Built Environment 2013 (ICABE2013)*, vol. 2013, no. November 2013, pp. 755–764, 2013.

[13] D. T. Doan, A. Ghaffarianhoseini, N. Naismith, T. Zhang, A. Ghaffarianhoseini, and J. Tookey, "A critical comparison of green building rating systems," *Build Environ*, vol. 123, pp. 243–260, 2017, doi: 10.1016/j.buildenv.2017.07.007.

[14] R. L. H. Chiu, "Social equity in housing in the Hong Kong special administrative region: A social sustainability perspective," *Sustain Dev*, vol. 10, no. 3, pp. 155–162, 2002, doi: 10.1002/sd.186.

2. A Study on Strength of Alkali-Activated Binders from Different Waste Materials

Nahan M, Niranjana S Mavelil, Silesh B Kamath and Deepak B*
Department of Civil Engineering, Government Engineering College Thrissur, Kerala, India
*Corresponding author: deepakb@gectcr.ac.in

ABSTRACT: The significant environmental impact resulting from the excessive use of natural resources and CO_2 emissions during cement production has prompted a growing emphasis on eco-friendly and sustainable construction materials. The management of construction and demolition waste poses an additional challenge. Geopolymers, which are amorphous to semi-crystalline alumina silicate polymers formed through an inorganic polycondensation reaction between solid alumina silicate and highly concentrated alkali hydroxide or silicate solution, have emerged as potential alternatives to ordinary cement. This study investigates the feasibility of alkali-activated geopolymer binders derived from various industrial by-products and construction and demolition wastes. Materials such as ground granulated blast furnace slag (GGBS), fly ash (FA), clay brick powder (CBP), Mangalore tiles powder (MTP), and rice husk ash (RHA) were activated with alkali to evaluate their potential. Mortar cube specimens, prepared using FA, GGBS, CBP, MTP, and RHA-based geopolymer binders, with a fixed concentration of sodium hydroxide, a specific ratio of sodium silicate to sodium hydroxide solution, and an alkaline activator to binder ratio, were subjected to ambient temperature curing and tested for compressive strength. Among the different mixes, GGBS and FA in combination with GGBS, exhibited the highest strength, thereby confirming their potential to completely replace cement. Additionally, MTP, in combination with GGBS-based geopolymer mix, also demonstrated sufficient strength.

KEYWORDS: Cement Replacement, Alkali Activation, Geopolymer, C&D Waste

1. INTRODUCTION

Concrete is the most widely used construction material owing to its low cost, ease of applicability, versatility, and reliability with yearly consumption levels approaching 30 billion tonnes. However, concrete production has a large environmental cost given the fact that the majority of the individual constituents forming this material are not eco-friendly and sustainable to manufacture. The most distinctive example of such constituents is Portland cement (PC) which is the main binder for traditional concrete and requires highly energy-intensive stages of production resulting in release of nearly 1 ton of CO_2 per ton of PC manufactured.

Furthermore, the production of cement clinker consumes large amounts of energy, averaging 850 kcal/kg, and enormous amounts of raw materials, averaging 1.7 tons for every 1 ton of clinker produced. Due to concerns about the tremendous amounts of concrete and related cement production worldwide, the latest research efforts into the development of more eco-friendly binders have been intensified. In this regard, "alkali-activated materials" (AAMs), including those referred to as "geopolymers" can be regarded as a major advancement toward the realization of greener binders (Provis 2017; Duxson et al., 2017).

2. ALKALI-ACTIVATED MATERIALS

Alkali activation is the generic term which is applied to the reaction of a solid alumina–silicate (termed the "precursor") under alkaline conditions (induced by the "alkali activator"), to produce a hardened binder which is based on a combination of hydrous alkali–alumino–silicate. There are two main pathways by which alkali-activated binders can be produced, either a one-part mix (dry powder combined with water) or a two-part mix (liquid activator) system. The two-part mix type is probably the main pathway that will be followed in the initial deployment of alkali-activation in most markets, and the majority of the products that are already in the market are produced in this manner. As alkali-activators, the most commonly used compounds are MOH and $M_2O \cdot rSiO_2$, where M is either Na or K.

3. MATERIALS AND EXPERIMENTAL DETAILS

3.1. Materials

3.1.1. Fly ash

Fly ash was collected from Malabar Cements Ltd, Walayar, Kerala. It was then sieved to a particle size of less than 90 μm using IS sieve. Fly ash based geopolymer concrete has been widely studied (Ryu et al., 2023). Temperature cured Fly Ash based geopolymer bricks has been previously explored by Balakrishnan et al (2019).

3.1.2. Ground granulated blast furnace slag

Ground granulated blast furnace slag (GGBS) was collected from Chennai. It was then sieved to a particle size less than 90 μm. GGBS were added to each binder sample by 20% weight of the dry powder to enable ambient temperature curing which is otherwise impossible to achieve. Since GGBS contains more amount of CaO, hydration reaction occurs with the addition of water which is exothermic, thus producing enough heat required for curing.

3.1.3. Mangalore tile powder

Mangalore tile (MT, also Mangalorean tiles) is a type of tile native to the south of India. They are a rich source of alumina and silica and thus are pozzolanic (Usha et al., 2016). MT was collected from locally available C&D wastes. It was then crushed and powdered using ball mill and then sieved using 90 μm sieve.

3.1.4. Clay brick powder

The waste clay brick is a kind of C&D waste with alkali activation potential (Hwang et al (2018); Yıldırım et al (2021); Villaquirán-Caicedo et al (2021). Clay bricks were collected from C&D wastes. It was then crushed and powdered using ball mill and then sieved using IS 90-micron sieve.

3.1.5. Rice husk ash

Rice husk ash (RHA) is a by-product of agriculture and is generated in rice mills. RHA was collected from Chennai. It was then sieved through IS 90-micron sieve.

Figure 1 shows the raw materials used for the experiment.

Figure 1. FA, GGBS, MT powder, RCB powder, and RHA samples.

The chemical compositions of each precursor material obtained by scanning electron microscopy energy dispersive X-ray (SEM-EDAX) analysis are shown in Table 1.

Table 1. Mineralogical composition of materials.

Material	% weight				
	SiO_2	K_2O	Al_2O_3	CaO	FeO
Fly Ash	98.54	1.46	-	-	-
GGBS	30.74	1.01	19.55	38.69	-
MT powder	50.67	2.37	33.39	-	11.79
RCB powder	46.55	1.21	27.07	1.29	-
RHA	65.05	1.09	22.91	-	10.95

3.1.6. Alkali activators

A mixture of sodium hydroxide (SH) and sodium silicate (SS) solutions was used as the alkali-activator. The alkali activator was prepared by dissolving 99% pure NaOH flakes in distilled water and mixed with SS gel in specified ratio. The concentration of the alkaline activator plays a crucial role in the geopolymerization reaction. The ratio SS/SH solution and molarity of sodium hydroxide were fixed.

3.2. Experimental details

3.2.1. Specimen preparation

All the raw materials were powdered using ball mill. The specimens were oven dried and sieved through 90 μm sieve. Only finer particles (<4.75 mm) were taken for M-sand in the experiment. Keeping molarity of NaOH as 11, alkali activator to binder ratio (A/B) as 0.8 and SS to SH ratio as 2.5, the alkali activator was prepared by dissolving NaOH pellets in distilled water and mixed with SS gel at least one day prior to the casting of geopolymer specimens.

At ambient temperature, geopolymerization reaction of the raw materials (FA, MT, RCB, and RHA) were extremely slow, which was enhanced by mechanical activation or addition of GGBS to the materials.

The mixes used for the study were as follows: FA8020 (80% FA and 20% GGBS), BR8020(80% CBP and 20% GGBS), GG100 (100% GGBS), MT8020 (80% MT Powder and 20% GGBS), and RH8020 (80% RHA and 20% GGBS).

The raw materials and the fine aggregates (M-sand) in the ratio 1:3 were first dry mixed manually in a container. After mixing, the alkali activator was added and mixing continued for another 4–5 minutes. The geopolymer mortar specimens (70.6 × 70.6 × 70.6 mm) were casted and then cured under ambient temperature. Unlike ordinary cement mortars, geopolymers require temperature curing, and not water curing. Experimental programme and the mix proportions are shown in Table 2. Figure 2 shows the casted cubes.

Table 2. Mix proportions for various raw materials of geopolymer mortar per cube.

SlNo.	Mix ID	Molarity (M)	A/B ratio	SS/SH ratio	Sample raw material (g)	GGBS (g)	Sand (g)	SS gel (g)	NaOH flakes (g)	Water (g)
1	GG100	11	0.8	2.5	-	200	600	114.3	13.97	31
2	FA8020	11	0.8	2.5	160	40	600	114.3	13.97	31
3	BR8020	11	0.8	2.5	160	40	600	114.3	13.97	31
4	MT8020	11	0.8	2.5	160	40	600	114.3	13.97	31
5	RH8020	11	0.8	2.5	160	40	600	114.3	13.97	31

Figure 2. *Cast specimens FA8020, GG100, MT8020, BR8020, and RH8020.*

3.2.2. Testing

The precursor materials were tested for SEM/EDX and X-ray diffraction (XRD). The hydrometer analysis was carried out to get the particle distribution.

The compressive strength test was carried out using compressive testing machine to evaluate the strength development of the specimens after specified days as per IS: 4031 (part 6).

4. RESULTS AND DISCUSSIONS

4.1. Results

Table 3 shows the combined test results of various geopolymer mortar cubes.

Table 3. Test results of various geopolymer mortar cubes in ambient temperature curing.

Specimen ID	FA8020	GG100	MT8020	BR8020	RH8020
7 day compressive strength (MPa)	25.63	34.93	18.7	4.8	0.53
28 day compressive strength (MPa)	48.4	57.1	24.56	4.97	3.73

4.2. Discussions

4.2.1. Influence of particle size distribution of precursors on compressive strength

The finer GGBS, FA, and MTP gave good results whereas coarser CBP and RHA gave unsatisfactory results. An increase in compressive strength was observed for increase in fineness of the precursor resulting in higher reactivity. Figure 3 agrees the above explanation.

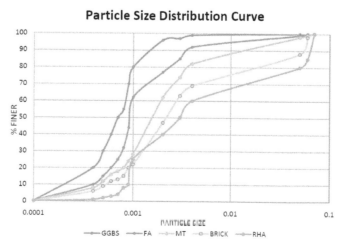

Figure 3. *Fineness of all precursors.*

4.2.2. Influence of amorphousness of precursors on compressive strength

The mineral contents of raw binder were determined by the quantitative X-ray diffraction (QXRD) method. Figure 4.1–4.5 shows the XRD spectrum of all raw materials. As per the studies, when the amorphousness of a material increases, the reactivity also increases (Konig et al., 2021; Costa et al., 2021). Similarly, when the crystallinity increases, the reactivity decreases [12]. It is seen (Figure 4.1 and 4.2) that the spectra of GGBS and fly ash have broad bands, respectively. These broad bands indicate the presence of more amorphous phase in the raw binder materials which contributes to the strength. Similarly, the spectra of MTP, CBP, and RHA have narrow bands, respectively. These peaks indicate the presence of crystalline phase in the raw binder materials which imparts lower strength.

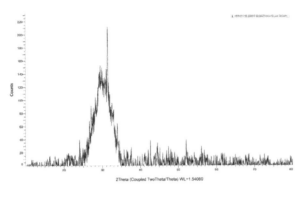

Figure 4.1. XRD of GGBS.   *Figure 4.2. XRD of FA.*

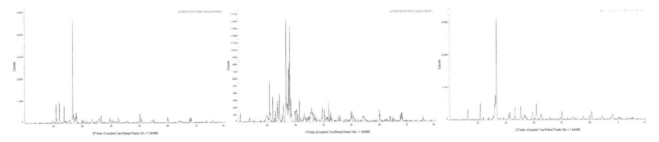

Figure 4.3–4.5. XRD of MT powder, brick powder, RHA, respectively.

4.2.3. Influence of silica–alumina composition of precursors on compressive strength

An increase in strength was observed for increase in the amount of alumina and silica content of the precursors. This is due to the increased geopolymerization reaction since hardening of alkali-activated binders were achieved by dissolving the aluminum and silicon components of precursors by alkaline activators (Madani et al., 2021). Also, the activation of powder becomes quicker due to the increase of solubility of alumina–silicate which produce higher content of the reaction products.

4.2.4. Cost analysis

Cost of alkali activator per cube was calculated to be Rs. 5.44. The total cost/cube of different mixes including the cost of alkali activator were: GG100—Rs.6.84, FA8020—Rs. 6.52, BR8020—Rs. 6.52, MT8020—Rs. 6.68, and RH8020—Rs.6.68. The cost of a cement mortar cube was Rs. 2.2.The cost of geopolymer mortar cube is thus higher when compared to normal cement mortar cube. Purchasing of materials in bulk quantity will reduce the price by a considerable range, especially of the chemicals which contributes to a major portion of the price difference.

5. CONCLUSIONS

Based on the results obtained from the experiment, the following conclusions can be made.

- Geopolymer binders have considerable potential to be used as an alternative to ordinary cement as a construction material in several applications.
- Water curing is not required for geopolymer binder, thus saving of water.
- The production of ready mixed alkali-activated binder can be achieved by the addition of 20% GGBS to each binder sample which enables ambient temperature curing instead of elevated temperature curing.
- In this study, the highest strength was obtained for GG100 (57.1 MPa) due to the presence of high amount of CaO (38.69 %) along with alumina and silica in GGBS and the fineness of the GGBS particles.
- High strength was obtained for FA8020 (48.4 MPa) due to the presence of very high amount of silica (98.54 %) in fly ash and fineness of the particles.
- Strength was reduced for MT8020 (24.56 MPa) and BR8020 (4.97 MPa) possibly due to very low amorphousness of the MT and brick powder and lower amount of alumina.
- Lowest strength was obtained for RH8020 (3.73 MPa) since the RHA particles are coarser and less reactive than the particles of other precursors.

REFERENCES

[1] Chao-Lung Hwang, Mitiku Damtie Yehualaw, Duy-Hai Vo, and Trong-Phuoc Huynh, "Development of high-strength alkali-activated pastes containing high volumes of waste brick and ceramic powders," *Const Build Mat*, 2019.

[2] Gum Sung Ryu, Young Bok Lee, Kyung Taek Koh, and Young Soo Chung, "The mechanical properties of fly ash-based geopolymer concrete with alkaline activators," *Const Build Mat*, 2013.

[3] Gürkan Yıldırım, Anıl Kul, Emircan Ozçelikci, Mustafa S¸ Ahmaran, Alper Aldemir, Diogo Figueira, and Ashraf Ashour, "Development of alkali-activated binders from recycled mixed masonry-originated waste," *J Build Eng*, 2021.

[4] H. Madani, A.A. Ramezanianpour, M. Shahbazinia, and E. Ahmadi, "Geopolymer bricks made from less active waste materials," *Const Build Mat*, 2020.

[5] John L. Provis, "Alkali-activated materials," *Const Build Mat*, 2017.

[6] Katja Konig, Katja Traven, Majda Pavlin, and Vilma Ducman, "Evaluation of locally available amorphous waste materials as a source for alternative alkali activators" *Ceram Int*, 2021.

[7] Leonardo Martins Costa, Natanael Geraldo Silva Almeida, Manuel Houmard, Paulo Roberto Cetlin, Guilherme Jorge Brigolini Silva, and Maria Teresa Paulino Aguilar, "Influence of the addition of amorphous and crystalline silica on the structural properties of metakaolin-based geopolymers," *Appl Clay Sci*, 2021.

[8] Mónica A. Villaquirán-Caicedo and Ruby Mejía de Gutiérrez, "Comparison of different activators for alkaline activation of construction and demolition wastes," *Const Build Mat*, 2021.

[9] Niveditha Balakrishnan, S. Usha, and Ponny K. Thomas, "Fly ash based geopolymer bricks: A sustainable construction material, "*Const Build Mat*, 2019.

[10] P. Duxson, A. Fernández-Jiménez, J. L. Provis, G. C. Lukey, A. Palomo, and J. S. J. Van Deventer, "Geopolymer technology: The current state of the art," *Const Build Mat*, 2017.

[11] S Usha Deepa G. Nairb, and Subha Vishnudas, "Feasibility study of geopolymer binder from terracotta roof tile waste," *Constr Build Mat*, 2016.

3. Feasibility of Residual Rice Husk Ash as a Source Material in Alkali-Activated Binder

Vinu Vijayan* and Deepa G Nair

Department of Civil Engineering, Cochin University of Science and Technology

*Corresponding author: vinuvijayan@cusat.ac.in

ABSTRACT: The energy-intensive and resource-depletive manufacturing process of cement in the construction industry is a major contributor to global warming and environmental degradation. Geopolymers are introduced as a sustainable alternative to cement. The objective of this study is to explore the feasibility of residual rice husk ash (RHA), the industrial waste from rice milling industries as source material for geopolymer composite. The results indicate the feasibility of residual RHA-based geopolymer composite under elevated curing conditions as suitable for building applications of low-strength requirements.

KEYWORDS: Geopolymerisation, RRHA, Alkali Activator, Molarity, A/B Ratio

1. INTRODUCTION

The cement demand is increasing every year. Global cement production in the year 2021 was reported as 4300 metric tons (IEA). At the same time, the production of cement is highly resource-intensive and environmentally polluting. Carbon dioxide released during the cement manufacturing process contributes approximately 5–7% of the total anthropogenic carbon dioxide emissions (Jindal, 2019). The significance of geopolymer comes in this context as a sustainable alternative to cement. Geopolymers are produced by the alkali activation of alumino–silicate materials. It has better mechanical and durability properties than ordinary Portland cement binders.

Residual rice husk ash (RHA) is an industrial waste left out after using rice husk as a fuel. These materials are usually found dumped on the premises of rice milling industries in Kerala and contributes to a serious environmental concern. Previous researchers have proved the potential of RHA produced under controlled temperatures in the geopolymerization process (Ahsan and Hossain, 2018; Jamil et al., 2013). However, the feasibility of residual RHA is not yet explored. The objective of this study is to investigate the potential of residual RHA as a source material for geopolymer composite.

2. EXPERIMENTAL RESEARCH

2.1. Material characterization

Materials used in this research are residual RHA, manufactured sand, and sodium hydroxide. Sodium hydroxide (NaOH) pellets purchased from a local supplier mixed with specific quantities of water were used as alkaline activators in this research.

Residual RHA (RRHA): RRHA used for this research was collected from Pavizham Rice Mills (Perumbavoor, Kerala). The sample was subjected to sieve analysis, X-ray diffraction analysis, and EDAX analysis. The particle size distribution of RRHA (Figure 1) indicates the fineness of the sample (71.2% of sample were finer than 300-micron size) and X-ray diffraction (Figure 2) indicates the presence of amorphous silica (68.4%). The results of the EDAX analysis presented in Table 1 show the presence of total silica as 89.93%.

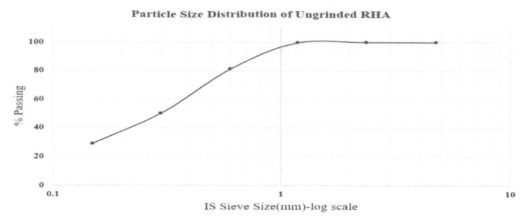

Figure 1. *Particle size distribution curve of RRHA.*

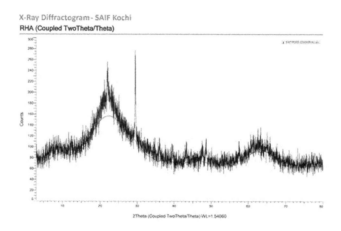

Figure 2. *X-ray diffraction pattern of RRHA.*

Table 1. Chemical composition of RRHA.

Composition	SiO_2	K_2O	CaO	MgO
Percentage by weight	89.93	3.31	5.08	1.68

Manufactured Sand: Manufactured sand passing through a 2.36 mm IS sieve and retained on a 75µm IS sieve was used as the fine aggregate (conforming to Zone II). The physical properties of fine aggregate were accomplished as per IS: 2386-1963.

2.2. Optimization of parameters

RRHA and manufactured sand mixed in the proportion 1:3 for the preparation of geopolymer mortars were initially prepared by varying the concentrations of alkaline activator solution (12, 13, 14 M) to investigate the influence of molarity of alkaline activator in the geopolymerization process. The other two parameters were kept constant (A/B-0.6, curing temperature 80°C for 24 hours) during the initial stage. The values for the selected parameters were decided based on previous research (Aliabdo et al., 2016; Bidwe and Hamane, 2015; Hamidi et al., 2016; Kaur et al., 2018; Kim et al., 2014). Alkaline activators of different molarities were prepared one day prior to the casting of the specimens. Additional water was also added for improving the workability of the mix. However, the water-to-binder ratio of 0.35 was maintained in all the mixes. Mortar cube specimens of 70.6 mm size were prepared and covered with aluminum foil and subjected to elevated temperature curing (80°C for 24 hours). Compressive strength tests were conducted as per IS:4031 (Part 6) (1988) after specific periods of ambient curing. Results are presented in Table 2. Based on the results, 13 M was selected as the optimum concentration for further investigations.

Table 2. Compressive strength with variation in molarities of NaOH.

Sl. no.	Molarity (M)	A/B ratio	Compressive strength (MPa)		
			3rd day	7th day	28th day
1.	12	0.6	4.11	5.84	6.72
2.	13	0.6	10.88	13.14	14.88
3.	14	0.6	11.18	13.04	14.12

The research was further continued to investigate the influence of the alkaline activator-to-binder ratio. A/B ratio was varied (0.4, 0.6, and 0.8), molarity was kept constant (13 M) and elevated curing condition was adopted (80°C for 24 hours) for the preparation of mortar samples and tests were conducted. The results of compressive strength tests are presented in Table 3. Alkaline activator-to-binder ratio of 0.6 was selected as the optimized value owing to the results of compressive strength tests.

Table 3. Compressive strength with variation in A/B ratio.

Sl. no.	A/B ratio	Molarity	Compressive strength (MPa)		
			3rd day	7th day	28th day
1.	0.4	13	3.72	3.85	6.24
2.	0.6	13	10.88	13.14	14.88
3.	0.8	13	4.53	12.67	14.63

To investigate the influence of fine particles in the geopolymerization process, residual rice husk particles were subjected to a ball mill for 30 minutes of grinding and sieved through a 90 μm IS sieve. Mortar specimens were prepared using this sample of RRHA with optimized conditions (13 M, A/B-0.6). Another set of specimens was prepared for investigating the effect of ambient curing. Compressive strength tests were conducted and the results are presented in Table 4.

Table 4. Compressive strength of ground RRHA samples in different curing conditions.

Type of RHA	Curing condition	Compressive strength (MPa) 28th day
Ground RHA	Ambient	10.53
	Elevated temperature	24.6

3. RESULTS AND DISCUSSIONS

This section presents the discussions of the results of this study to verify the influence of concentration of alkaline activator, alkali activator-to-binder ratio, grinding, and curing conditions.

3.1. Influence of concentration of alkaline activator

Figure 3 shows the variation of compressive strength results of RRHA-based geopolymer mortar using different molarities for different periods of testing.

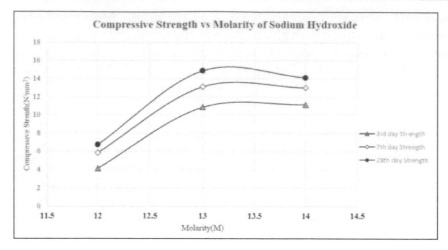

Figure 3. Variation of compressive strength with molarity at different days of testing.

The results show that the strength increases with increasing molarity up to an optimum (13 M) and then decreases with an additional increment of 1 M (i.e., at 14 M) a decreasing trend in strength was observed. This can be due to the presence of excess OH-ions hindering the geopolymerization process. The maximum compressive strength of the specimen at 13 M indicates the formation of an active geopolymer gel network due to the presence of active silica released during the dissolution of RRHA.

3.2. Influence of alkaline activator-to-binder ratio

Figure 4 shows the variation of compressive strengths for different A/B ratios (0.4, 0.6, 0.8) and maximum compressive strength was observed with A/B-0.6.

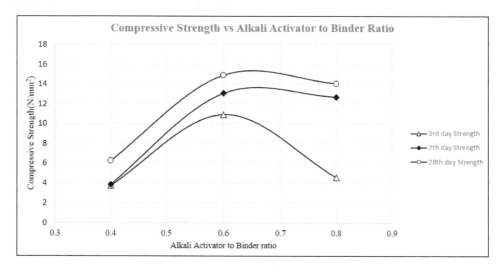

Figure 4. Variation of compressive strength with A/B ratio at different days of testing.

At low A/B ratios (below 0.6), the mixture was less workable and hence due to the non-homogeneity of the mix, a reduction in compressive strength was observed. Lower water content at a low A/B ratio causes difficulty for dissolved species to diffuse, leading to a reduction in polycondensation. However, at A/B ratio of 0.6, better workability was noticed and maximum compressive strength was obtained. This can be due to the complete dissolution of binder material enabling the formation of geopolymer gel. At a higher A/B ratio (above 0.6), reduction in compressive strength was noticed. This may be due to lower binder content and higher alkaline activator content causing chances of the formation of air bubbles in the matrix leading to a reduction in strength.

The strength may increase with increasing the value of A/B ratio and molarity of NaOH due to higher rate of geopolymerization and resulting dense microstructure. However, the variation depends upon the chemical composition of source material. The optimum values of A/B (0.6) and molarity (13 M) of the proposed geopolymer composite correspond to low amorphous silica content of RRHA.

3.3. Influence of grinding

Figure 5 shows the influence of grinding on the compressive strength of the proposed RRHA-based geopolymer composite. A significant increase in strength (77.6%) can be noticed after grinding. This can be attributed to the improved fineness and increased surface area of the source material due to grinding. Hence, improved reactivity, faster rate of geopolymerization, and increase in compressive strength.

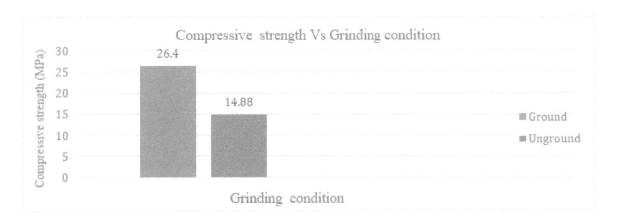

Figure 5. *Compressive strength (28th day) versus grinding condition.*

3.4. Feasibility of ambient curing condition

Figure 6 shows the variation in compressive strength of the test specimens cured under ambient and elevated conditions for different periods of testing. The compressive strength test results of RRHA specimens prepared under elevated curing conditions were higher than ambient curing conditions in all periods of testing. Higher strength of geopolymer specimens cured at elevated temperatures can be attributed to the high rate of polymerization.

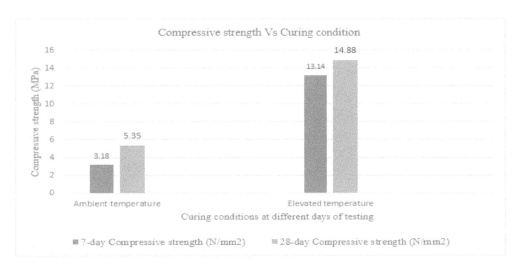

Figure 6. *Compressive strength with ambient and elevated curing at different ages of testing.*

4. CONCLUSIONS

The feasibility of residual RHA as source material for geopolymer composite was established. Optimized parameters for the proposed geopolymer binder were identified (concentration of NaOH—13 M and alkali activator-to-binder ratio—0.6). The curing condition adopted was 80°C for 24 hours for the entire study.

The findings of the feasibility study conducted are as follows:

- Geopolymer composite developed using residual RHA (as received) as source material can be used for secondary building applications (14.88 MPa).
- Geopolymer composite using residual RHA (after grinding and sieving) having finer particles (<90 µm) as source material can be used for structural applications with low strength requirements (26.4 MPa).

Feasibility of ambient curing in geopolymerization of residual RHA needs further investigations to make it suitable for sustainable construction.

REFERENCES

[1] Ahsan, M. B., and Hossain, Z., "Supplemental use of rice husk ash (RHA) as a cementitious material in the concrete industry," *Const Build Mat*, vol. 178, pp. 1–9, 2018.

[2] Aliabdo, A. A., Elmoaty, A. E. M. A., Salem, H. A., "Effect of water addition, plasticizer and an alkaline solution constitution on fly ash based geopolymer concrete performance," *Const Build Mat*, vol. 121, pp. 694–703, 2016.

[3] Bidwe, S. S., and Hamane, A. A., "Effect of different molarities of sodium hydroxide solution on the strength of geopolymer concrete," *Am J Eng Res*, vol. 4, no. 3, pp. 139–145, 2015.

[4] Hamidi, R. M., Man, Z., Azizli, K. A., "Concentration of NaOH and the effect on the properties of fly ash based geopolymer," *Procedia Eng*, vol. 148, pp. 189–193, 2016.

[5] IEA. https://www.iea.org/reports/cement (accessed: October 25, 2022)

[6] IS: 4031 (Part 6), "Methods of physical tests for hydraulic cement," Bureau of Indian Standards, 1988.

[7] Jamil, M., Kaish, A. B. M. A., Raman, S. N., and Zain, M. F. M., "Pozzolanic contribution of rice husk ash in cementitious system," *Const Build Mat*, vol. 47, pp. 588–593, 2013.

[8] Jindal, B. B., "Investigations on the properties of geopolymer mortar and concrete with mineral admixtures: A review," *Const Build Mat*, vol. 227, pp. 116644, 2019.

[9] Kaur, K., Singh, J., and Kaur, M., "Compressive strength of rice husk ash based geopolymer: The effect of alkaline activator," *Const Build Mat*, vol. 169, pp. 188–192, 2018.

[10] Kim, Y. Y., Lee, B. J., Saraswathy, V., and Kwon, S. J., "Strength and durability performance of alkali-activated rice husk ash geopolymer mortar," *Sci World J*, 2014.

4. Feasibility Study of Using White Mortar as Plaster Mortar in Building Construction

Subha Vishnudas[1,*] and Midhun Joseph[2]

[1]Faculty of Civil Engineering, School of Engineering, Cochin University of Science and Technology, Cochin, Kerala

[2]Section Engineer, Kerala Rail Development Corporation Ltd, Trivandrum, Kerala

[*]Corresponding author: v.subha@cusat.ac.in

ABSTRACT: When a building's aesthetic beauty and structural soundness are combined, it will earn a reputation as a good building. Ordinary Portland cement (OPC), which is typically used in concrete, masonry, and plastering works, and grey or brown aggregates (both coarse and fine) produce finished surfaces that are predominantly grey in color. When finalizing the aesthetic elements of a building, designers and architects often choose white color. For imparting white color to grey-colored OPC works, multiple courses of white putty and white emulsion have to be applied. These additional painting works will be time-consuming, which is generally crucial while constructing a building, and will also involve added expenses. The need for further painting will be eliminated if white mortar is used in masonry and plastering works, and the plastered surface will be ready for commissioning after the curing period. Even though White Portland Cement (WPC) is more expensive than OPC, the economic benefits would be estimated by comparing them to the expenses involved with painting masonry and plastering works that are using OPC. This study examines the viability of combining white marble sand as the fine aggregate and WPC as the binder to develop white mortar that can be utilized in building construction.

KEYWORDS: Marble Sand, White Mortar, Properties of Mortar

1. INTRODUCTION

White cement is a wonderful option for aesthetic and architectural masonry and plastering work because it offers a neutral tinting base and consistent color effects. Every color, from vivid and pastel hues to pure whites, may be utilized to paint it. The extent of post-construction aesthetic work can be decreased if a spectacular and alluring structure can be built with white mortar. In various proportions of 5–50% with an increment of 5%, Fadia and Aziz (2018) tested the properties of cement mortar after replacing a portion of the cement with white cement. Cube specimens were cast in a 1:2:5 ratio with a varying water-to-cement ratio ranging from 25 to 50%. The test result revealed that a 25% replacement of white cement for OPC resulted in the highest compressive strength. A study on a newer kind of white cement with TiO_2 that has photocatalytic activities was conducted by Luigi Cassar et al. (2003). This cement enables concrete to maintain its aesthetic qualities over time while also contributing to the elimination of harmful pollutants from the urban environment. Within 30 h of casting, the utilization of white cement in high performance concrete resulted in a high compressive strength of 41 N/mm² being observed. Accordingly, it was expected that using white cement in mortar mixtures would result in strong mortar that was comparable to that produced by ordinary Portland cement (OPC). Khayalia et al. (2017) investigated the strength characteristics of a lean mortar mixture including marble waste as a sand replacement (1:6 binder/sand ratio). Crushed marble waste was substituted for fine aggregate in amounts ranging from 25 to 100% with increments of 25% by sand volume. It was observed that at 28 days of curing period, the compressive strength increased with the marble powder substitution level increased and reduced at 100%

Chapter 4 DOI- 10.1201/9781032657271-4

replacement. Cement mortar with a cement/sand ratio of 1:3 and waste marble powder was investigated by Rai et al. (2011) to determine its compressive strength. The quantity of waste marble powder utilized to replace fine aggregate ranged from 5 to 20% by weight. A number of researchers in their investigations have found that marble sand can be utilized as fine aggregate for the production of white mortar (Dave et al., 2016; Hebhoub et al., 2011; Raghvendra and Trivedi, 2017; Sakalkale et al., 2014; Tamrakar et al., 2018). Waste marble dust added to concrete in place of fine aggregate increases its compressive, tensile, and flexural strength. When fine aggregate is substituted for other materials in the production of white concrete and mortar, the flexural strength and compressive strength have been enhanced by up to 50%.

2. MATERIALS AND METHODS

Binder and fine aggregates are the essential ingredients for the experimental study. The mortar that used natural sand as the fine aggregate and OPC as the binder was regarded as the control mix. Test specimens were prepared to investigate (i) the characteristics of mortar using white Portland cement (WPC) as the binder and natural sand as the fine aggregate to evaluate the strength of WPC mortar and (ii) the properties of mortar utilizing WPC and marble sand to assess the effect of using white marble sand as a fine aggregate replacement. On the theory that if leaner mixes like 1:5 are used, the content of white cement will be less and it may affect the intensity of whiteness of the mortar, 1:3 and 1:4 mix proportions for mortar were chosen to develop a white colored mortar as a plaster mortar that may not necessitate further finishing works over it. Use of the white-colored marble sand produced mortar with a high degree of whiteness. Findings were assessed in accordance with ASTM C-348 and C-349 as well as Indian Standard codes such IS 8042-1989 (2009), IS 2116-1980 (2002), IS 2185-2005 (2005), IS 2250-1981 (2000), and IS 383-2016 (2016).

2.1. Materials

The materials used for the experimental study include:

- Ordinary Portland Cement 53 grade
- River sand
- White Portland Cement 53 or higher grade
- Marble sand (white)

2.1.1. Material characterization

2.1.1.1. Cement

Test for consistency, fineness, setting time, and compressive strength were conducted to evaluate the properties of WPC and OPC.

Table 1. Properties of OPC and WPC.

Property	OPC	WPC
Consistency (%)	31	35
Fineness (%)	1	0.50
Initial setting time (min)	65	70
Final setting time (min)	<600	<600
Compressive strength		
3rd day (N/mm²)	27	30
7th day (N/mm²)	35	40
28th day (N/mm²)	54	57

The properties of OPC and WPC are presented in Table 1. It shows that the consistency of WPC is greater than OPC, which might be due to its greater fineness. Initial setting time of WPC is also found to be greater than OPC. WPC is found to have greater compressive strength than OPC.

2.1.1.2. River sand and marble sand

To study the characteristics of river sand and marble sand, sieve analysis and specific gravity tests were carried out. The specific gravity of river sand was found to be 2.67. The sieve analysis was conducted for river sand as per IS 383-2016 (2016). The fineness modulus of river sand was found to be 2.48 and hence it is classified as fine sand.

2.1.1.3. Marble sand

The specific gravity of marble sand was found to be 2.65 comparable to river sand. The sieve analysis was conducted as per IS 383-2016 (2016). The fineness modulus of marble sand was found to be 3.15. The marble sand is found to be coarse sand compared to river sand.

2.2. Experimental setup

2.2.1. Specimen preparation

- As per ASTM C-348 (2021), rectangular prisms of 160 mm × 40 mm × 40 mm were cast for determining flexural strength.
- Portions of the mortar prisms tested in flexure according to this test method are used for the determination of compressive strength in accordance with test method ASTM C-349 (2008).
- 70.6 mm IS standard cubes are used for determining water absorption and sorptivity.

Three sets of specimen for each test were prepared to study and analyze various properties of the mortar. The flexural strength and compressive strengths were tested using the same specimen, rectangular prisms, as per ASTM C-348 (2021) and C-349 (2008). Water absorption and sorptivity were tested using mortar cube specimens as per IS codes. The details are furnished in the succeeding sections. The cubes for testing the water absorption and sorptivity are prepared as per IS codes (IS 2185-2005, 2005). The properties of cements are tested as per IS 4031(6)-1988 (2005) (OPC) and IS 8042-1989 (2009) (WPC). The fine aggregates are tested as per IS 383-1970 (2016) and IS 2116-1980 (2002).

3. RESULTS AND DISCUSSIONS

The results of strength and durability tests are discussed in this section.

- Ordinary grey mortar using OPC (53 grade) and river sand in mix ratios of 1:3 and 1: 4 were tested to study the mechanical properties.
- White mortar using WPC and river sand in mix ratios of 1:3 and 1:4 were tested and compared with OPC mortar properties.
- The river sand in white mortar was replaced by white marble sand in 100% and analyzed different properties.

3.1. Mechanical properties

The tests were conducted using OPC/WPC and river sand/marble sand in the ratio 1:3 and 1:4. Tables 2 and 3 show the properties of 1:3 and 1:4 mortar mix, respectively. Figure 1 shows the test setup for flexural strength.

Table 2. Mechanical properties of 1:3 mortar.

Sl. No	Particulars	Flexure strength (N/mm²)			Compressive strength (N/mm²)		
		Day 3	Day 7	Day 28	Day 3	Day 7	Day 28
1	1:3 (OPC:River sand)	2.74	2.95	3.57	12.06	13.24	21.00
2	1:3 (WPC:River sand)	3.78	4.04	4.45	20.97	24.52	26.01
3	1:3 (WPC:Marble sand)	4.90	5.25	5.38	21.73	24.80	26.97

Figure 1. *Testing for flexural strength of OPC and WPC mortar prisms.*

Table 3. Mechanical properties of 1:4 mortar.

Sl. No	Particulars	Flexure strength (N/mm²)			Compressive strength (N/mm²)		
		Day 3	Day 7	Day 28	Day 3	Day 7	Day 28
1	1:4 (OPC:River sand)	2.25	2.64	3.31	7.72	9.64	19.95
2	1:4 (WPC:River sand)	2.49	2.85	3.78	9.28	12.06	20.70
3	1:4 (WPC:Marble sand)	2.54	2.92	3.93	9.47	12.45	21.84

As per IS 2250 (1981) (2000), in Table 1, the highest grade of masonry mortars shown is MM 7.5, that is, 7.5 N/mm² compressive strength at 28 days. As per clause 7.1.4 of above mentioned IS code, the minimum grade of masonry mortar to be used in buildings is MM 3. In this study, the compressive strength obtained for 1:3 white mortar is 26.97 N/mm² and that of 1:4 mortar is 21.84 N/mm², respectively.

3.2. Water absorption

The water absorption test was conducted using 70.6 mm cubes as per IS specifications. Cubes after curing period were kept in oven and after attaining constant weight, dry weights were taken. Then, cubes were immersed in water for 24 h. After 24 h, wet weights were taken and the difference in weight is expressed as % of dry weight to denote water absorption of the mortar cubes.

Table 4. Water absorption of mortars.

Sl. No	Particulars	Water absorption (%)	
		1:3	1:4
1	OPC:River sand	7.56	9.67
2	WPC:River sand	8.14	9.94
3	WPC:Marble sand	7.54	8.62

The water absorption is found to be minimal for the mix using WPC and white marble sand. The water absorption is found to be minimal for the mix using WPC and white marble sand. This might be attributed to the fineness of WPC and filler effect of marble sand which results in filling up of pores and hence reduction in water absorption. As per clause 9.5 of IS 2185 (Part 1) (2005) for hollow and solid concrete blocks, the water absorption being the average of three units, when determined shall not be more than 10% by mass. In Table 4, values obtained for mortar mixes of WPC and river sand as well as marble sand are within the limits specified in the above-mentioned IS code.

3.3. Sorptivity

The test procedure for sorptivity was carried out as per ASTM C 1585 (2021). The unit of sorptivity is $mm/s^{1/2}$. Table 5 shows the sorptivity values of the test specimen.

Table 5. Sorptivity of mortars.

Sl. No	Particulars	Sorptivity ($mm/s^{1/2}$)	
		1:3	1:4
1	OPC:River sand	0.0217	0.0633
2	WPC:River sand	0.0204	0.0381
3	WPC:Marble sand	0.0201	0.0207

As per ASTM C 1585 (2021), sorptivity value less than 0.099 $mm/sec^{(1/2)}$ is considered as excellent. Accordingly, the sorptivity value for all the three mortar mixes comes under the excellent grade. Still sorptivity value is found to be higher in conventional OPC and river sand mix in the case of 1:4 mortar whereas in the case of 1:3 mortar, the differences between the various mixes were found to be marginal. Hence the effect is more pronounced in the case of leaner mixes. Sorptivity value is comparatively less in WPC and marble sand mix. The reason behind this may be due to the fineness of WPC compared to OPC. Marble sand would be imparting a filler effect also which results in the reduction of pores and hence sorptivity.

4. COST–BENEFIT ANALYSIS OF WHITE MORTAR

It is well known in the construction industry that the cost of WPC is 60% or higher than OPC which makes it a less preferable choice for massive works like concrete, masonry, and plastering. Now white cement is mostly used as a whitewash over the plastered concrete or masonry surface to give an aesthetic look to the surface. White cement is used in some stone masonry and marble flooring works also.

The rate analysis of white mortar and convention OPC mortar (1:3 and 1:4) has been carried out as per CPWD Analysis of Rates for Delhi, Volume 1 published by Central Public Works Department, Government of India (2019) which can be used all over India with the application of local cost index. It is estimated that the cost of one cum of 1:3 ratio white mortar is 82% more than the OPC mortar and the cost of one cum of 1:4 ratio white mortar is 74% more than the OPC mortar. Since the quantity of labor required for masonry and plastering works will be the same for both OPC and White cement mortars, the same has not been considered in the rate comparison. The cost abstract is shown in Table 6.

Table 6. Cost abstract of OPC and WPC mortars.

Sl. No:	Mix ratio	Unit	Cost of OPC mortar	Cost of WPC mortar
1	1:3	cum	Rs.4183	Rs.7622
2	1:4	cum	Rs.3529	Rs.6154

The expenditure of white painting works to be done on OPC-plastered surfaces has to be analyzed to review the real economic impact of WPC mortar as painting work is not required for surfaces plastered with WPC mortar. For the purpose of analysis, the mix ratio 1:4 is considered. The standard thickness for wall plastering is 12 mm and considering the above, the cost of OPC mortar required for plastering 1 m² is found to be Rs. 42.35 and that of WPC mortar for same work is found to be Rs.73.85.

Table 7. Cost of painting works as per DAR 2019.

Sl. No	DAR code	Specification	Unit	Total cost (material + labor) (Rs.)
1	13.43.1	Application of thinable cement primer	m²	60
2	13.60.1	Wall painting with acrylic emulsion paint—two coats	m²	128.65

Again, the cost of one coat of primer and two coats of emulsion including labor is considered for rate analysis. Putty work is not considered since it gives more smoothness and finishing to the surface which cannot be achieved by bare plastering with white cement mortar. Table 7 shows the cost required for painting 1 m² area. By analyzing above details, the cost comparison of OPC mortar plastering with white painting works and WPC mortar works are shown in Table 8.

Table 8. Cost comparison of OPC and WPC mortar works.

Sl. No	Particulars	Unit	Cost of mortar (Rs.)	Cost of primer (Rs.)	Cost of emulsion (Rs.)	Total cost (Rs.)
1	OPC mortar-plastered wall	m²	42.35	60	128.65	231
2	WPC mortar-plastered wall	m²	73.85	Nil	Nil	73.85

As per Table 8, the cost of OPC mortar plastering, cement primer, and white emulsion works is found to be 68% more than the WPC mortar plastering at the time of construction itself and there again requires frequent painting over the years. The application of cement primer and white emulsion is not necessary for surfaces plastered with WPC mortar as the surface will look in white color even after plastering. Hence it can be inferred that using white cement and white mortar for plastering work will bring aesthetical and economic advantages. Also, the time required for painting works can be saved which is a critical factor in the completion of projects, especially commercial projects. The white mortar in a ratio 1:4, that is, one part of white cement and four parts of marble sand can be used for the plastering of both ceiling and walls.

5. CONCLUSIONS

From the experimental investigation with WPC and white marble sand in plaster mortar, following conclusions were drawn:

- As per IS 2250 (1981) (2000), in Table 1, the highest grade of masonry mortars shown is MM 7.5, that is, 7.5 N/mm² compressive strength at 28 days. As per the IS code, the minimum grade of masonry mortar to be used in buildings is MM 3. The compressive strength of 1:3 white mortar is 26.97 N/mm² and that of 1:4 mortar is 21.84 N/mm². This shows that white mortar can be used in building works.
- As per IS 2185 (Part 1):2005 (2005), the highest compressive strength of Grade A hollow concrete block is 15 N/mm² and that of solid concrete block is 5 N/mm². The high compressive strength and flexural strength of white mortar enlighten the possibility of using white cement and marble aggregates for concrete works, particularly for making white concrete blocks so that a structure can be white color in its construction stage itself.

- Water absorption and sorptivity are found comparatively less for mortar mix of WPC and marble sand.
- By means of cost analysis, it has been found that the use of white mortar will bring economic advantages by saving the cost and time for painting works.

REFERENCES

[1] Aalok D. Sakalkale, G. D Dhawale, and R. S. Kedar, "Experimental study on use of waste marble dust in concrete," Int J Eng Res Appl, vol. 4, no. 10, pp. 44–50, 2014.

[2] Amit Tamrakar, Gourav Saxena, and Tushar Saxena, "The consequences of waste marble dust on the mechanical properties of concrete," Int Res J Eng Technol, vol. 5, no. 9, pp. 285–288, 2018.

[3] ASTM C 1585-04, *Standard Test Method for Measurement of Rate of Absorption of Water by Hydraulic Cement Concretes*, American Society for Testing and Materials, 2021.

[4] ASTM C 348-02, *Standard Test Method for Flexural Strength of Hydraulic Cement Mortar*, American Society for Testing and Materials, 2021.

[5] ASTM C 349-08, *Standard Test Method for Compressive Strength of Hydraulic Cement Mortar (Using Portions of Prisms Broken in Flexure)*, American Society for Testing and Materials, 2008.

[6] CPWD, *Analysis of Rates for Delhi*, Volume 1, New Delhi: Central Public Works Department, Government of India, 2019.

[7] Dave, N., Misra, A. K., Srivastava, A., and Kaushik, S. K, "Experimental analysis of strength and durability properties of quaternary cement binder and mortar," Const Build Mat, vol. 107, pp. 117–124, 2016.

[8] Hebhoub, H., Aoun, H., Belachia, M., Houari, H., Ghorbel, E., "Use of marble aggregates in concrete," Const Build Mat, vol. 25, pp. 1167–1171, 2011.

[9] IS:2116-1980, *Specification for Sand for Masonry Mortars*, 1st revision, New Delhi: Bureau of Indian Standards, Reaffirmed 2002.

[10] IS:2185-2005, *Concrete Masonry Units (Part 1 Hollow and Solid Concrete Blocks)*, 3rd revision, New Delhi: : Bureau of Indian Standards, 2005.

[11] IS:2250-1981, *Code of Practise for Preparation and Use of Masonry Mortars*, 1st revision, New Delhi: Bureau of Indian Standards, Reaffirmed 2000.

[12] IS:383-2016, *Coarse and Fine Aggregate for Concrete-specification*, 3rd revision, New Delhi: Bureau of Indian Standards, 2016.

[13] IS:4031-1988, *Method of Physical Test for Hydraulic Cement (Part 7)*, 1st revision, New Delhi: Bureau of Indian Standards, Reaffirmed 2005.

[14] IS:8042-1989, *White Portland Cement-Specification*, 2nd revision, New Delhi: Bureau of Indian Standards, Reaffirmed 2009.

[15] Khyaliya, R. K, Kabeer, K. S. A, and Vyas, A. K., "Evaluation of strength and durability of lean mortar mixes containing marble waste," Const Build Mat, vol. 147, pp. 598–607, 2017.

[16] Klak, Fadya and Abdulla, Aziz, "Compressive strength of cement mortar with white cement and limestone," Int J Eng Technol, vol. 7, 2018. doi: 10.14419/ijet.v7i4.37.23614.

[17] Luigi Cassar, Carmine Pepe, Giampietro Tognon, Gian Luca Guerrini, and Rossano Amadelli, "White cement for architectural concrete, possessing Photocatalytic properties," 11th International Congress on the Chemistry of Cement, Durban, 2003.

[18] Raghvendra, and M.K Trivedi, "Partial replacement of cement with marble dust powder in concrete," Int J Res Appl Sci Eng Technol, vol. 5, no. 5, pp. 712–717, 2017.

[19] Rai, B., Naushad, K. H., Abhishek, K., Rushad, T. S., and Duggal, S. K., "Influence of marble powder/granules in Concrete mix," Int J Civil Struct Eng, vol. 1, no. 4, pp. 827–834, 2011.

5. Supply Chain Management in Construction Industry

Shakeeb Shakeel* and Preetpal Singh

Department of Civil Engineering, Rayat Bahra University, Mohali, Punjab, India

Corresponding author: Shakeebshakeel98@gmail.com, Preetpal.17966@rayatbahrauniversity.edu.in

ABSTRACT: There can be a wide range of issues in the building supply chain caused by a disturbance in the normal flow of supply. Various resources, such as consumer and management data, are transferred during the construction process (main contractor related). This study may be the first of its kind to analyze how construction supply chain flows affect project performance in residential dwelling construction projects in terms of delays and potential influence. In this study, the major goal is to investigate the impact of supply chain management (SCM) on construction project performance. Competitiveness in the building supply chain is critical. Businesses benefit from construction SCM by becoming more competitive, generating more income, and having more control over the numerous project components and factors. The development store network, supply chain concerns, and the advantages of a coordinated inventory network in the development sector are all covered in this article. When it comes to boosting a company's performance, SCM has received a lot of attention. Because construction is such a significant social and economic activity in every country, increasing the competitiveness of construction firms and the building industry is a major goal of SCM strategies. Construction projects of all sizes, scopes, and complexity may be found all over the world. Construction enterprises must incorporate a wide range of supply chains and markets when offering a solution to a customer or client, which goes against the advice of prior literature that generic supply chains be simple and linear. That is why the goal of the thesis is to review and integrate data from earlier research on supply chains in construction to serve as a roadmap for future investigation.

KEYWORDS: Supply Chain Management, Construction Management, Stakeholders, CPM, PERT

1. INTRODUCTION

Recently supply chain has become a major subject of management research and manufacturing theory. Supply chain has been defined as the network of organizations which are involved through upstream and downstream linkages, in the different processes and activities that produce value in the form of services and products in the hands of the ultimate customer (de Sousa Jabbour et al., 2011). The readiness of supply chain in integrated material procurement is one of the important indicators to define the performance of contractors (Segerstedt and Olofsson, 2010). Contractor will be called competent if they can order the material and build without any delays. Supply chain management (SCM) has been widely regarded as an effective and efficient management measure and strategy to improve the performance of the construction industry, which has suffered from high fragmentation, large waste, poor productivity, cost and time overruns, and conflicts and disputes for many years (Duboi and Gadde, 2000). Currently, SCM is in its developing stage. In this current scenario to avoid the conflicts and competition among the suppliers, a proper SCM system is needed. It mainly consists of number of participants and is complex in its nature. The construction sector players include engineers, contractors, suppliers, and clients (Wibowo and Sholeh, 2015). They have major roles in establishing and developing SCM and collaboration. There are many challenges that are faced by the construction industry in India, and the important challenge among them is the improper material supply chain in construction. Each and every product that reaches an end user is the cumulative effort of multiple organizations (Saad et al., 2002). These organizations refer collectively to as a supply chain (Mistry et al., 2003). Supply chain network of organizations and business processes for procuring materials, transforming raw materials into finished products, and distributing the finished products to the customers (Singh, 2011). SCM is an integration of suppliers, distributors, and customer logistics into one cohesive process.

2. SCM: THE ROLE IN CONSTRUCTION

Introduced in the manufacturing industry SCM is a vital tool in controlling business processes in a defined and systematic way to improve quality, save time, and increase profit (Vrijhoef and Koskela, 2000). On the other hand, SCM processes in the construction industry are partially adapted and dispersed.

Dr Ghaith Al-Werikat holds a Doctorate in Engineering Project Management, Coventry University.

According to their research findings (Balwani et al., 2010) claimed that the relationship among clients, suppliers, and contractors is mainly about purchasing and production planning. This reflects that SCM in the UK construction sector is partially adopted (Pettersson and Segerstedt, 2013) contributed that SCM roles in construction are vital and important. They claimed that, within construction, there are four major roles of SCM. These roles can be identified based on the industry concerns, whether it is the entire supply chain, the construction site, or both as displayed in Figure 1.

Role 1: focus on the interface between the supply chain and the construction site

Role 2: focus on the supply chain

Role 3: focus on transferring activities from the construction site to the supply chain

Role 4: focus on the integrated management of the supply chain and the construction site

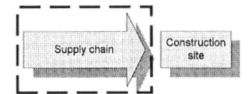
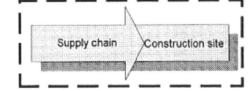

3. RESEARCH METHODS OVERVIEW

The logic, process, purpose, and outcome are all parts of a study, according to Collis and Hussy (2013). According to the rationale, we can tell if the research is going from being very particular to being more general. Inductive and deductive approaches to research fall under this umbrella. The second category is the procedure for gathering and analyzing information, which discloses the process. Qualitative or quantitative research falls under this umbrella. The rationale for performing the study falls under the purview of the third category, "purpose." Classifications under that umbrella include analytic, descriptive, predictive, and exploratory. Fourth, the research outcome tells whether the research is likely to make a significant advance to knowledge or resolve a specific problem. Basic and applied research fall under this umbrella category. The classifications and categories of research are laid out in Table 1.

Research type	Classification base
Deductive or inductive research	The rationale for the study
Inquiry-based or survey-based research	The study's methodology
Inquiry-based or exploratory research	The goal of the investigation.
The investigation's primary objective	Findings from the study

4. THE ADOPTED RESEARCH METHODS

As part of our investigation, we will be determining our aims and objectives. Preliminary investigations, surveys, and simulation models were used as depicted in Figure 2.

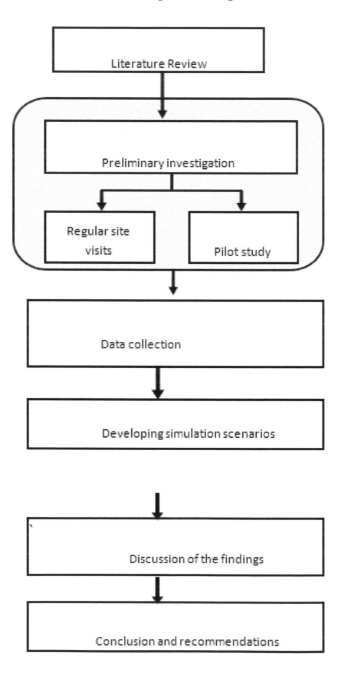

5. SCM BENEFITS IN CONSTRUCTION

Modern-day approaches to methods involving the procurement process are the movement toward an integrated supply chain. This method enables parties within the supply chain to have goal congruence, through alignment of objectives, which in turn provides the client with added value. Traditionally, the relationship between companies and the client was by means of contracts only, with predetermined prices and predefined specifications. Clients were not heavily involved, contractors were not motivated to work in the interest of the client and often had one contract with the client and a separate contract with designers. The movement toward an integrated supply chain enables provision of SCM to be wholly incorporated. Benefits of integrated supply chains for companies are as follows:

- Cost reduction and waste reduction
- Risk reduction, with a more certain final project cost
- Value for client
- Enables long-term planning
- Ongoing business or repeat business (with client)

Ultimately, clients and end users gain by being party to an industry which facilitates users' needs. Projects are completed in a timely manner to cost and defects are minimalized, resulting in customer satisfaction and indeed a greater confidence in the construction industry. Moreover, Erikson (2010) contributed that an integrated supply chain in construction offers more control and aids in cost reduction.

6. CONCLUSION

SCM in the construction business is a challenging endeavor. As a part of this collective, a large number of people contribute their time and efforts to the project. Many factors influence the construction industry's supply chain, and the conceptual corrective remedies for each study are determined by the research's convenience.

By applying SCM, the construction industry can obtain more project control, increase profits, and reduce the amount of time and resources required. In terms of size, the CSC has two major divisions: the material and construction supply chains. In order to improve the efficiency of the decision-making process, it is necessary to integrate the construction and material supply chains. SCM in the construction industry is beset by a slew of problems, including inadequate logistical planning, a lack of strategic links with suppliers, resistance to change, and poor lines of communication. An effective supply chain relies on long-term partnerships, open channels of communication, and the pooled knowledge of all participants. This knowledge can only be obtained through working together.

Studies and discussions were done on SCM in various fields based on the journals collected. SCM is a challenging task in the construction industry and is complex in their structure. It is composed of a large number of participants who work together in the project in a temporary manner. Different factors affecting supply chain in construction fields are to be identified and the conceptual remedial measures for each study are going to be suggested according to its convenience. The details regarding the topic are to be collected by questionnaire survey with the help of internal and external personalities involved in the system. Different methods are suggested by the authors for the ranking of the factors in the management system like RII technique, Fussy logic method for maximization and minimization techniques, etc. The most effective questionnaire was the Likert scale method in which each and every one can respond according to his/her will.

This study was conducted on a very small-sized project and the saving was found to be around 6.25%. However in bigger projects, if the supply chain is well established the profits could be much more. In every project, a well-established supply chain is required and the proper planning of the project should be done a few months in advance, taking into account all the stakeholders.

REFERENCES

[1] Agung Wibowo, M., and Sholeh, MohNur, "The analysis of supply chain performance measurement at construction project," *Procedia Eng*, vol. 125, pp. 25–31, 2015.

[2] Ana Beatriz Lopes de Sousa Jabbour, Alceu Gomes Alves Filho, Adriana Backx Noronha Viana, and Charbel Jos Chiappetta Jabbour, "Factors affecting the adoption of supply chain management practices: Evidence from the Brazilian electro-electronic sector," *IIMB Manag Rev*, vol. 23, pp. 208–222, 2011.

[3] Anders Segerstedt, and Thomas Olofsson, "Supply chains in the construction industry," *Supply Chain Manag*, vol. 15, no. 5, pp. 347–353, 2010.

[4] Anna Duboi, and Lars-Erik Gadde, "Supply strategy and network elects purchasing behavior in the construction industry," *Eur J Purch Supply Manag*, vol. 6, pp. 207–215, 2000.

[5] Annelie I. Pettersson and Anders Segerstedt, "To evaluate cost savings in a supply chain: Two examples from Ericsson in the telecom industry," *Oper Supply Chain Manag*, vol. 6, no. 3, pp. 94–102, 2013.

[6] Mohammed Saad, Martyn Jones, and Peter James, "A review of the progress towards the adoption of supply chain management (SCM) relationships in construction," *Eur J Purch Supply Manag*, vol. 8, pp. 173–183, 2002.

[7] Mukesh Balwani, S., Hussain, S. A., Aquib Ansari, and Naseeruddin Haris, "Supply chain management in construction," *Int J Rec Innov Tren Comp Commun*, vol. 3, no. 2, 2010.

[8] Rajen B. Mistry, Vishal R Gajera, and Hiren A. Rathod, "Evaluation of factor affecting for supply chain in construction project," *Int J Adv Res Eng Sci Manag*. ISSN: 2394-1766.

[9] Rajesh K. Singh, "Developing the framework for coordination in supply chain of SMEs," *Bus Process Manag J*, vol. 17, no. 4.

[10] Ruben Vrijhoef, and Lauri Koskela, "The four roles of supply chain management in construction," *Eur J Purch Supply Manag*, vol. 6, pp. 169–178, 2000.

[11] Sonja Petrovic-Lazarevic, Margaret Matanda, and Russell Worthy, "Supply chain management in building and construction industry: Case of Australian residential sector," Monash University Business and Economics, Working Paper 21/06, pp. 1–8, 2006.

6. Experimental Study on Performance of Cold-Mix Asphalt Using Different Additives

Faisal Bashir* and Ajay Vikram

Department of Civil Engineering, Rayat Bahra University, Mohali, Punjab, India

Corresponding author: aryanfaisal36@gmail.com, Ajayvikram99151@gmail.com

ABSTRACT: Cold-mix asphalt (CMA) has been upgraded or modified in this study such that its mechanical qualities are comparable to hot-mix asphalt (HMA). Mechanical qualities including indirect tensile stiffness modulus, permanent deformation resistance, and stiffness that creeps can be improved by adding cement, lime, or waste materials (by-products). Using CMA instead of HMA, which uses heat in both the production and laying processes, is a major advance in the field of asphalt paving. CMA has similar engineering features to HMA. Instead of HMA, a high-performance CMA combination will be developed and refined for use as a surface track (HMA). Increased structural resistance to distress and increased pavement life expectancy can also be attributed to these alterations. The technology of hot mix has witnessed significant advancements in numerous research programs. As can be seen in India, technology of cold mix is much behind the times in terms of research and practical use. It is the primary rationale for selecting the CMA as the focus of the current study. In fact, compared to HMA, this has both environmental and economic advantages.

KEYWORDS: Cold-Mix Asphalt, Marshal Stability, OPC, Lime, Indirect Tensile Test

1. INTRODUCTION

There are a number of road-building ideas and events that focus on HMA-coated pavements. For numerous years, HMA, a more traditional method of road construction, has met efficiency standards on a structural level. The criteria that technology of hot mix commonly uses are: drying of the binder and aggregate, mixing, grinding of the tack, mixing, and compaction, all of which are done at increased temperatures between 120 and 165°C. Environmental deterioration, increased carbon dioxide in the air, low manufacturing efficiency, low rain and cold weather laying job capacity, restricted construction duration within a year and oxidative binder stiffening are some of the drawbacks of this high utilization in pavement systems (AI-Busaltan et al., 2012). A suitable alternative to the HMA would therefore be ideal. Rural road initiatives with large economic burdens are also underperforming in India's northern and northeastern regions, such as J&K. It is hard to apply technologies of hot mixture in these steep sites because of topographical and environmental constraints, areas with high runoff, and wooded zones. Cold mix equipment is the only way to keep research going, no matter what the terrain is like. HMA's work is not jeopardized even by the most harsh weather conditions. In this post, the developments thus far have been examined to see if we can produce a substitute for the traditional techniques of installing Asphalt. Numerous studies have examined the effectiveness of cold mix samples to find out if they can replace traditional hot mixes in areas with topographical limits (Asi and Assaad, 2005; Asphalt Institute Manual Series, 1997; ASTM D, 2007). This study emphasizes the importance of innovation in renewable energy development. Studies have shown that progress in renewable technology is needed in the area of road construction, and this has led to a focus on the development of "natural roads" as a strategy for long-term development (Bueno et al., 2003).

Chapter 6 DOI- 10.1201/9781032657271-6

Alternative materials and methods, such as those derived from renewable resources, are used to their full potential in green construction. Reducing traffic and infrastructure uses the negative influence on a healthy and prosperous society is the goal here. The design, installation, and use of materials in the road-engineering industry will lead to environmentally sustainable infrastructure (green infrastructure). With the goal of going green, cold mix technology is making its way into road construction.

Mix architecture based on cold mixing binders with the proper IRC aggregates without the requirement for heating is referred as cold mix technology in practice.

By adding prewetting liquid to aggregate, followed by emulsion, and then creating the mix, all processes in the emulsion-based cold mixing method are completed at room temperature. In addition, field studies have demonstrated that mixing may be easily produced utilizing HMA to lay down using the same processes. Work-friendly, too.

2. METHODOLOGY

2.1. Aggregates

Cold-asphalt mix is produced by mixing unheated mineral aggregate with either emulsified bitumen or foamed bitumen. Unlike hot-mix asphalt (HMA), CMA does not require any heating of aggregate which makes it economical and relatively pollution free (no objectionable fumes or odors). Production of CMA does not require high investment in equipment, which makes it economical. It is also suitable for use in remote areas. CMAs can be used both for initial construction (100% virgin mixes) and for recycling asphalt pavements.

2.2. Filler

Crusher dust or stone dust is a compactable, economical packing material used for stabilizing surfaces. Crusher dust is also known as blue metal, cracker, or rock dust. It is the material leftover when making crushed rock. While rocks are going through the crusher, tiny pieces, and dust particles are left behind. This dust is recycled and becomes a valuable product with many practical applications instead of being thrown out as waste.

2.3. Binder

In its most common form, asphalt binder is simply the residue from petroleum refining. To achieve the necessary properties for paving purposes, binder must be produced from a carefully chosen crude oil blend, and processed to an appropriate grade. Asphalt binders are most commonly graded by their physical properties and these properties are significantly influenced by temperature. An asphalt binder's physical properties directly describe how it will perform as a constituent in asphalt concrete (AC) pavement. Although asphalt binder penetration and viscosity grading is still commonly used worldwide, new binder tests and specifications were developed to more accurately characterize asphalt binders for use in AC pavements. An emulsion of bitumen acquired from a reliable source was used to treat the samples in this research project. The asphalt's residue concentration was determined to be 64.9%, according to the study.

2.4. Cement

In the present study, the effect of cement on cold mix design has been studied. Various cold mix samples are prepared using bitumen emulsion, aggregates, and some another admixture. Comparative analysis is carried out to know the effect of cement as admixture on the properties such as Marshall stability value flow value, air voids of bituminous mixture samples containing different amount of admixture. It is one of the most widely used types of Portland cement. The name Portland Cement was named by Joseph Aspdin in 1824 due to its similarity in color and its quality when it hardens like Portland stone.

2.5. Tests on materials

The materials were to be put through the following tests to check whether they are perfect for the utilization for the purpose of the study.

2.5.1. Tests on aggregates

The following tests were carried out on aggregates:

1. Aggregate impact test.
2. Los Angeles abrasion Test.
3. Aggregate crushing value test.
4. Shape tests.
5. Soundness test.

2.5.2. Tests on binder

1. Residue by evaporation.
2. Determination of penetration on a sample of residue.
3. Determination of ductility on a portion of residue.
4. Determination of specific gravity by pycnometer method.

2.5.3. Tests on CMA

1. Marshall stability test.

3. RESULTS AND DISCUSSIONS

Many researchers in industrialized countries such as the USA, the UK, and Australia examined the qualities of cold mix to assure its ongoing usage as a base course or surface treatment material. This involved looking for ways to improve cold mix for these uses, as well as other upkeep tasks like pothole filling (Borhan et al., 2009).

Several researchers have assessed CMA's performance and come to the conclusion that it differs from HMA in certain ways. Although the bitumen and aggregates are heated to between 140 and 160°C in HMA, they are combined at room temperature irrespective of any heating process in CMA. Due to the lower CO_2 emissions of CMA compared to HMA, it is considered greener. CMA testing, in terms of indirect tensile strength tests, were found to be significantly lower than the 800 kPa minimum standard value for HMA, according to some researchers. A CMA's weaker mechanical characteristics and strength early in its life, combined with its higher permeability and lower stability, make it inferior to hot mixes in general. HMA, on the other hand, reaches its full potential in a few of weeks.

3.1. Aggregate test results

The following results Table 1 were obtained after conducting the various tests on the aggregates using proper guidelines for performing the tests on the aggregates:

Table 1. Aggregate test results.

Property	Test method	Test result (%)
Impact Value	The methods carried out are quoted in the sub-section 2.5.1	14.4
Crushing Value		13.01
Los Angles Abrasion Value		18
Flakiness Index		18.84
Elongation Index		21.4
Water absorption		0.13

3.2. Marshall test results

The samples were pressed down using the Marshall method in accordance with the aggregate gradation to achieve cold mixing after the coarse aggregate, fine aggregate, and crusher dust were mixed as filler material. It was decided to use the empirical formula to determine the initial residual asphalt content value for this particular cold mix gradation. At optimum residual asphalt content, other cold mix factors were also examined. The air void, Voids in mineral aggregates, stability loss, and flow values were all 6.4%, 8.28%, 14.7%, and 3.2 mm, respectively.

3.2.1. Effect of additives on design parameters

This was the primary goal of the research conducted, that is, to understand the effect of additives on design parameters (Brown and Needham, 2000). Additive mixing was performed independently in a range from 1 to 5% for a change of 1%. Figure 1&2

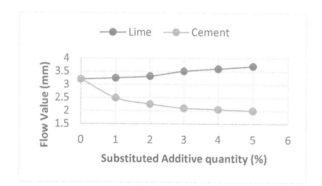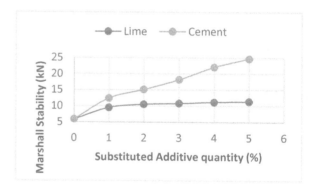

Figure 1. *Flow value and Marshall stability results of cold mix with additives.*

3.2.1. Effect of additives on indirect tensile strength

At the optimum emulsion content, the ITS test was performed on both unsoaked and soaked specimens (OEC). Higher ITS values were obtained for unsoaked specimens with the addition of cement, at 1.123 MPa, than for soaked specimens, at 0.893 MPa. Unsoaked specimens had an ITS of 0.989 MPa, while soaked samples had an ITS of 0.812 MPa. The ITS value demonstrates that cement can improve the mix's tensile strength.

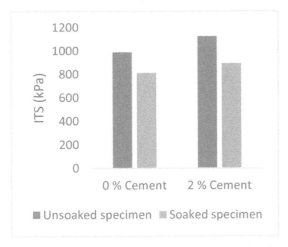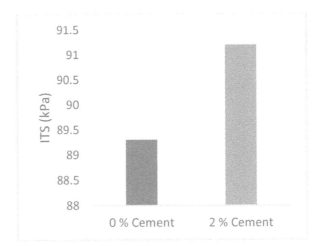

Figure 2. Indirect tensile strength for unsoaked and soaked specimen.

4. CONCLUSION

The stiffness modulus of cold mix following OPC modification was found to be comparable to hot mix. It takes time to achieve these engineering qualities but this can be remedied with the addition of OPC. Such combinations benefit from OPC's early strength development. Evaporation loss is responsible for this improvement, as the remaining water is utilized to slowly hydrate and harden concrete, so making OPC a binder in the mix with a proper water-cement ratio due to the evaporation. Some of the OPC hydrates and becomes a part of the binder, as demonstrated by scanning electron microscopy (SEM). Calcium chloride and hydrated lime were tested, however, the results showed that they had no effect on stiffness.

REFERENCES

[1] Al Buoaltan S., Al Nageim H., Atherton W., and Sharples G., "Mechanical properties of an upgrading cold-mix asphalt using waste materials," *J Mat Civil Eng*, vol. 24, no. 12, pp. 1484–1491, 2012.

[2] Asi I., and Assaad A., "Effect of Jordanian oil shale fly ash on asphalt mixes," *J Mat Civil Eng*, vol. 17, pp. 553–559, 2005.

[3] Asphalt Institute Manual Series No.14 (MS-14), *Asphalt Cold Mix Manual* (Third Edition), Lexington, KY, 1997.

[4] ASTM D 6931, *Indirect Tensile (IDT) Strength of Bituminous Mixtures*, Philadelphia, PA: American Society for Testing Materials, 2007.

[5] Benedito de S. Bueno, Wander R. da Silva, Dario C. de Lima, and Enivaldo Minnete, "Engineering properties of fiber reinforced cold asphalt mixes," *J Environ Eng*, vol. 129, no. 10, pp. 952–955, 2003.

[6] Borhan Muhamad Nazri, Suja Fatihah, Ismail Amiruddin, and Rahmat Riza Atiq O. K., "The effects of used cylinder oil on asphalt mixes," *Eur J Sci Res*, vol. 28, no. 3, pp. 398–411, 2003.

[7] Brown S. F., and Needham D., "A study of cement modified bitumen emulsion mixtures," *Proc AAPT*, pp. 69, 2000.

7. Impact of Lean Utilization on Operational Performance: A Study of Sri Lankan Textile and Apparel Industry

Ilangakoon S M [1] and Kumudinei Dissanayake [2,*]

[1]Faculty of Graduate Studies, University of Colombo, Sri Lanka

[2]Department of Management and Organization Studies, University of Colombo, Sri Lanka

[*]Corresponding author: kumudisa@mos.cmb.ac.lk

ABSTRACT: Apparel and textile manufacturers should enhance their operational performance in order to supply high-quality goods in shorter lead times. This requirement is met by the introduction of lean utilization. However, the mere implementation of lean does not result in improved operational performance. The present study examines the impact of lean utilization on operational performance and the effect of lean duration in the apparel industry in Sri Lanka. The study population comprises garment manufacturers who have adopted lean as their standard of operation for at least the main value-adding process for more than 1 year. The sample size is 30. Three main lean constructs and 16 lean practices have been tested in the study. Findings show a positive association between lean utilization and operational performance. Waste elimination is the highly utilized lean cluster in the apparel sector. The longer the lean production processes are in operation, the higher the operational performance in apparel and textile companies.

KEYWORDS: Lean Utilization, Lean Duration, Operational Performance, Textile and Apparel Industry, Sri Lanka

1. INTRODUCTION

With the global market becoming increasingly competitive, manufacturing companies are under immense pressure to achieve operational excellence and enhance their performance in order to lower costs and provide high-quality products in shorter lead times (Belekoukias et al., 2014). "Lean manufacturing principles and techniques have been widely used by manufacturing organizations to achieve these and gain a competitive advantage over their rivals" (Belekoukias et al., 2014, p. 5346). The literature shows that implementing lean production practices successfully results in a streamlined, high-quality system that produces products and services with increased productivity, lower costs, shorter lead times, and increased volume flexibility, all of which enhance an organization's performance (Shah and Ward, 2003). However, firms, on the other hand, frequently fail to gain the full benefits of lean tactics because they do not build the performance assessment measures required to assess the gains in effectiveness and efficiency (Gunasekaran et al., 2004). Lean principles are still being introduced, and the Sri Lankan garment sector has not yet fully benefited from them (Silva et al., 2011). The majority of apparel and textile manufacturers in Sri Lanka, still focus on basic garments where competition is primarily based on price, Sri Lanka faces stiff competition from other emerging nations of South and Southeast that have lower manufacturing costs. With its vast supply capacity and low manufacturing costs, China has also become a prominent power in the global garment sector. In comparison to Sri Lanka, these nations are ranked lower in terms of manufacturing costs. Therefore, it is clear that Sri Lanka's garment industry's operational performance is not at the highest level. The purpose of the project is to conduct an experiment to determine how lean utilization affects operational performance in Sri Lanka's textile sector and to objectively assess the connection between lean and operational performance.

Chapter 7 DOI- 10.1201/9781032657271-7

2. LITERATURE REVIEW

2.1. Lean practice in the manufacturing sector

One of the famous Japanese automobile manufacturers Toyota has climbed to a position of global importance in the automobile sector by changing the mass production system that has been transformed into a Toyota Production System (TPS; Black, 2007), which is now well-known all across the world as lean production. According to Belekoukias et al. (2014), in the modern global market being more competitive, manufacturing organizations are under massive stress to chase operational excellence and enhance performance to reduce costs of production and provide goods with higher quality in a shorter lead time. As identified by Garza-Reyes et al. (2012) (as cited in Belekoukias et al., 2014), "Lean manufacturing principles and techniques have been widely used by manufacturing organizations to achieve these and gain a competitive advantage over their rivals" (p. 5346). The goal is to make organizations more competitive in the market by decreasing costs through the elimination of non-value-added activities and inefficiencies in the process and increasing efficiency, the utilizing management approach can be considered as lean manufacturing (Belekoukias et al., 2014).

2.2. Lean utilization

Improvement in the effectiveness and efficiency of lean strategies to maximize the benefits is defined as lean utilization (Gunarathne and Kumarasiri, 2017). The continuous performance measurement matrices are useful tools for monitoring lean utilization continuously (Karim and Arif-Uz-Zaman, 2013). Industry-specific KPIs are essential for assessing the success of lean utilization in every organization (Gamage et al., 2012). However, the published research does not contain any such lean indicators specific to the clothing sector in Sri Lanka. A performance measurement system is essential for analyzing the leanness of operations and determining if the lean techniques and ideas have been successfully applied in the country's garment industry (Gunarathne and Kumarasiri, 2017).

2.3. Lean utilization and operational performance

Lean production system implementation produced better organizational results than other production manufacturing methods, such as flexible manufacturing systems and computer-integrated manufacturing systems, according to research (Anand and Kodali, 2009). According to Paneru (2011), the use of lean tools resulted in some significant benefits, including an 8% decrease in production cycle time, a 14% decrease in the number of operators needed to produce an equivalent number of garments, an 80% decrease in the amount of rework, a reduction in lead time from two days to one hour, and a decrease in the maximum WIP inventory from 500 to 1500 pieces to 100 pieces. In addition to these obvious advantages, operators' levels of multi-skilling and style changeover flexibility have also increased. Manufacturing performance is considerably improved by lean production techniques (Wickramasinghe and Wickramasinghe, 2017). The usage of lean tools depends on the nature of the industry, the plant size, and the technological capabilities of the country (Silva et al., 2011). The overall impact of lean techniques and tools on operational performance may still be seen as being unclear due to the type of study done (Belekoukias et al., 2014). Additionally, Sakakibara et al. (1997) indicated in their analysis that there was no discernible impact of just-in-time techniques on an organization's operational performance. However, just-in-time techniques are important for infrastructure.

2.4. Lean in the apparel industry of Sri Lanka

Being competitive in the region is a difficulty for the Sri Lankan garment industry (Gamage et al., 2012). The absence of a reliable raw material basis is the largest weakness of the Sri Lankan garment sector, according to Dheerasinghe (2009) when compared to other major rivals in Asia.

Sri Lanka is heavily reliant on imported raw materials and accessories. More than 70% of the raw materials utilized in this industry, as well as 70–90% of the accessories, are imported. Therefore, the cost of raw materials accounts for more than 70% of the cost of manufacturing. Lean was adopted in organizations, according to Gamage et al. (2012), in order to satisfy these criteria. Since the middle of the 2000s, Sri Lankan clothing exporters have realized that lean production offers them the chance to increase the effectiveness of their production processes (Wickramasinghe and Wickramasinghe, 2010). On the basis of the aforementioned research, we put up the following hypothesis: H1: Utilization of lean practices has an impact on the operational performance (OP) of apparel manufacturing firms.

2.5. The time duration of implemented lean practices

Lean duration is the amount of time that lean production is carried out (Wickramasinghe and Wickramasinghe, 2017). Since lean manufacturing can only be accomplished over time (Womack and Jones, 1996), it should not be utilized as a magic bullet to resolve immediate competitive issues. Lean duration should also be taken into account when analyzing the adoption of lean manufacturing techniques (Wickramasinghe and Wickramasinghe, 2017). On the basis of the literature study above, we suggest; H2: The longer the lean duration, the better the lean performance results will be. Lean duration moderates the link between the adoption of lean production processes and lean performance outcomes.

3. METHOD

The variables in the study have been framed as follows in order to meet the study goals and test the hypothesis.

3.1. The conceptual framework of the study

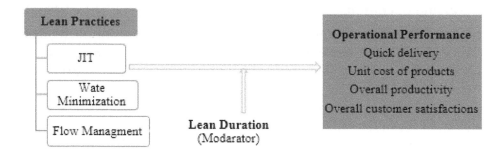

Figure 1. Conceptual framework.

3.2. Variables and measures

There were 16 lean practices identified and those were grouped into three main lean constructs called waste elimination (Pull production system/Kanban, Removing Bottlenecks, Poka Yoke, Waste elimination, Zero defects, Line Balancing), flow management (Reducing Production Lot Size, Focusing on one Supplier, Single Piece Flow, Cellular manufacturing), and just in time (Reducing Setup Time, Preventive Maintenance, Cycle Time Reduction, Reducing Inventory, New Process Technology and Equipment, Quick Change over Techniques). The leanness and operational performance were measured using a 5-point Likert scale ranging from 1 "strongly disagree" to 5 "strongly agree." Table 1 is shown the variables relevant to this study.

Table 1. Variables of linear regression model.

Independent variable	Lean construct	Dependent variable	OP measures
X1	Just in time	Y1	Quick delivery
X2	Waste elimination	Y2	The unit cost of products
X3	Flow management	Y3	Overall productivity
		Y4	Overall customer satisfaction

3.3. Research design

This study was carried out in a quantitative approach where data were gathered through a questionnaire survey and analyzed by using the SPSS statistical software. At the time of data collecting, it has done reference literature review and through the survey questionnaire. When collecting the required data 30 factories in Sri Lanka have been selected as the samples which Lean had implemented at least 1 year before. Statistical analysis was conducted, thus, correlation analysis was done in order to identify the relationship between implemented lean tools and operational performance, as well as the effect of lean duration.

3.4. Target population and sampling technique

Lean utilization in the apparel and textile industry has been well-noted among export-based firms (Wickramasinghe and Wickramasinghe, 2017) in the apparel and textile in the export sector. The study population was identified with the following criteria, that is, Sri Lankan medium to large-scale garment manufacturing enterprises who have adopted lean as their style of operation, at least in terms of the major value-adding process for more than 1 year. No special sampling procedure was used in this investigation, and the minimum sample size was determined to be the same as the market population. This was due to the restricted number of factories that met the aforementioned criteria, as well as the country's pandemic situation at the time of the study. Medium to large-scale 30 apparel manufacturing firms were derived. Data have been gathered from the relevant lean executives or senior management in the operational department who have worked for more than a 1-year period to ensure that the accuracy of data and knowledge provide reliable feedback to the questionnaire.

3.5. Data analysis

3.5.1. Reliability and validity

Internal reliability and validity are measured using one multivariate analysis called principal component analysis and Cronbach alpha value using the SPSS statistics software. The 16 total lean practices were subjected to the principal component analysis with varimax rotation from SPSS software to examine their unidimensional. This study's results show reliability and validity because factor loadings of the scales exceed 0.5 and Cronbach alpha values reported above 0.7 level.

3.6. Findings

Out of the main three lean constructs, the Waste elimination construct has the highest contribution to the model with a Beta value of 0.373 which is statistically the highest significance with a p-value of 0.000, which is less than the usual significance level of 0.05 ($p < 0.05$). Just in time and Flow management contributed to the model with a Beta value of 0.215 (significant level = .018) and 0.203

(significant level =.026) respectively with high significance. So, H1a is supported at the highest level whereas H1b and H1c are also supported in the middle. Moreover, the coefficient of linear regression predicted a positive or negative correlation between each independent variable and the dependent variable. Based on the results of the simple linear regression analysis, it is evident that all three main lean constructs have had a positive impact on OP. Therefore, it is proven that H1 is supported. The past literature (Wickramasinghe and Wickramasinghe, 2017) recommended going for hierarchical regression analysis when analyzing the moderator effect. In order to interpret the hierarchical regression results, it is required to look at the unstandardized regression coefficient (B) for each independent variable, the coefficient of determination (R2)) value, and the adjacent coefficient of determination (adjacent R2) value. Model 2, lean duration significantly impacts OP with adjacent R squared value 0.075 (p < 0.001). The interaction effects for each main three lean clusters and lean duration on OP are also significant with adjacent R squared value 0.175 at the 95% level of significance because the lower and upper levels of the confidence intervals did not include zero. Hence, this supports H2.

4. DISCUSSION, CONCLUSION, AND IMPLICATIONS

Based on empirical analysis, it is reasonable to draw the conclusion that Waste elimination techniques in big enterprises had a higher influence on OP in the Sri Lankan garment sector, even if the majority of international research claimed that Just in time was the most often used lean approach. In any case, the Sri Lankan garment sector today makes use of Just in time and Flow management techniques. Therefore, it could be wise to emphasize waste elimination techniques more in the garment business. Zero defects and line balance were not evaluated in the Sri Lankan context as a Waste elimination metric, but according to (Gunarathne and Kumarasiri, 2017), Waste elimination are the most used lean technique in the Sri Lankan apparel and textile sector. Line balance and zero faults are essential Waste elimination approach, as described in Paneru (2011). As a result, this investigation found zero defects, and line balancing greatly increases OP, making them the highly current focus of the Sri Lankan garment sector. The Sri Lankan garment sector today makes extensive use of Just in time strategies such as inventory reduction, cycle time reduction, and conducting preventive maintenance operations. Flow management practices have a significantly lower degree of influence in comparison to the other two primary lean components, which is consistent with the findings of Rahman et al. (2010) and Gunarathne and Kumarasiri (2017). According to Gunarathne and Kumarasiri (2017), Flow management techniques like concentrating on one supplier provide the smallest contributions to OP. Based on the empirical findings of this study, it can also be deduced that concentrating on a single supplier accomplishes little to advance the OP and has not yet gained much traction in Sri Lanka's garment sector. Despite the fact that Gunarathne and Kumarasiri (2017) conclude that single-piece flow does not improve Flow management, Paneru (2011) has promoted the use of single-piece flow as an Flow management practice. This study identified single-piece flow as highly contributing toward OP as an Flow management practice in the contemporary apparel industry in Sri Lanka, and further attention has been paid to the single-piece flow in this industry. Lean utilization has had a significant influence on OP, as objectively demonstrated around the world. Due to the fierce rivalry in the South Asian garment market, it has even been tested a few times in Sri Lanka using different criteria. A different cluster was chosen for the survey in this study in order to obtain a new perspective on the topic and analyze the influence of lean usage on OP using other criteria. The findings supported previous research and showed that lean usage had a favorable effect on OP. Not only in the Sri Lankan context but also worldwide, researchers (e.g., Rahman et al., 2010) had found a positive relationship between lean practices and the OP. The duration in lean production practices has been in operation as a standardized practice and is a critical parameter in lean production (Wickramasinghe and Wickramasinghe, 2017). Here, the moderator effect of the duration of lean techniques in the operation was found to enhance the OP. Thus, it was visible that the longer the lean production processes are in operation the higher the lean utilization We can recommend the top management of apparel firms to focus more on the Waste elimination lean practices in order to reduce the cost of production as well as to improve the OP. Not only the Waste elimination practice but also some of the Just in time and Flow management practices contribute greatly to improving the OP of apparel firms, as revealed in the study. For future research, it would be suggestible to expand the present study with different scopes and under similar circumstances, thus covering diverse levels of lean experiences and different lean practices. Further, it would be more reliable to conduct the empirical survey directly approaching the operational staff. Past research has been conducted by approaching the top-level management or the executive-level employees in the investigation of similar issues. As the apparel industry is a labor-intensive industry, it is good to investigate ground-level.

REFERENCES

[1]   Anand, G., and Kodali, R., "Selection of lean manufacturing system using the analytic network process – a case study," J Manuf Technol Manag, vol. 20, no. 2, pp. 258–289, 2009.

[2]   Belekoukias, I., Garza-Reyes, J. A., and Kumar, V., "The impact of lean methods and tools on the operational performance of manufacturing organizations," Int J Prod Res, vol. 52, pp. 5346–5366, 2014.

[3]   Black, J. T., "Design rules for implementing the Toyota Production System," Int J Prod Res, vol. 45, pp. 3639–3664, 2007.

[4]   Dheerasinghe, R., "Garment industry in Sri Lanka challenges, prospects and strategies," Staff Stud, vol. 33, pp. 33–72, 2009.

[5]   Gamage, J. R., Vilasini, P. P. G. N., Perera, H. S. C., and Wijenatha, L., "Impact of lean manufacturing on performance and organizational culture: A case study of an apparel manufacturer in Sri Lanka," A paper presented at the 3rd International Conference on Engineering, Project and Production Management (EPPM), Brighton, United Kingdom, 2012. http://www.ppml.url.tw/EPPM/conferences/2012/download/SESSON3_B/40%20E133a.pdf

[6]   Garza-Reyes, J. A., Oraifige, I., Soriano-Meier, H., Forrester, P. L., and Harmanto, D., "The development of a lean park homes production process using process flow and simulation methods," J Manuf Technol Manag, vol. 23, pp. 178–197, 2012.

[7]   Gunarathne, G. C. I., and Kumarasiri, W. D. C. K. T., "Impact of lean utilization on operational performance: A study of Sri Lankan textile and apparel industry," VJM, vol. 3, pp. 27–41, 2017.

[8]   Gunasekaran, A., Patel, C., and McGaughey, R. E., "A framework for supply chain performance measurement," Int J Prod Econ, vol. 87, pp. 333–347, 2004.

[9]   Karim, A., and Arif-Uz-Zaman, K., "A methodology for effective implementation of lean strategies and its performance evaluation in manufacturing organizations," Bus Proc Manag J, vol. 19, pp. 169–196, 2013.

[10]   Kelegama, S., and Epaarachchi, R., "Productivity, competitiveness and job quality in garment industry in Sri Lanka," A Discussion Paper, ILO, 2001.

[11]   Paneru, N., "Implementation of lean manufacturing tools in garment manufacturing process focusing sewing section of men's shirt," Masters Thesis, Oulu University of Applied Sciences, 2011. https://www.theseus.fi/bitstream/handle/10024/34405/Paneru_Naresh.pdf

[12]   Rahman, S., Laosirihongthong, T., and Sohal, S. A., "Impact of lean strategy on operational performance: a study of Thai manufacturing companies," J Manuf Technol Manag, vol. 21, pp. 839–852, 2010.

[13]   Sakakibara, S., Flynn, B. B., Schroeder, R. G., and Morris, W. T., "The impact of just in-time manufacturing and its infrastructure on manufacturing performance," Manag Sci, vol. 43, pp. 1246–1257, 1997.

[14]   Shah, R., and Ward, P. T., "Defining and designing measures of lean production," J Oper Manag, vol. 25, no. 4, pp. 785–805, 2003. doi: 10.1016/j.jom.2007.01.019

[15] Silva, S. K. P. N., Perera, H. S. C., and Samarasinghe, G. D., "Viability of lean manufacturing tools and techniques in the apparel industry in Sri Lanka," Appl Mech Mat, vol. 110–116, pp. 4013–4022, 2011.

[16] Wickramasinghe, D., and Wickramasinghe, V., "Perceived organizational support, job involvement and turnover intention in lean production in Sri Lanka," Int J Adv Manuf Technol, vol. 55, pp. 817–830, 2010.

[17] Wickramasinghe, G. L. D., and Wickramasinghe, V., "Implementation of lean production practices and manufacturing performance: The role of lean duration," J Manuf Technol Manag, vol. 28, pp. 531–550, 2017.

[18] Womack, J. P., and Jones, D. T., Lean Thinking: Banish Waste and Create Wealth in Your Corporation, New York, NY: Simon and Schuster.

8. Strength and Permeability Characteristics of Cement-Treated Subbase Layer

Saranya Ullas[*] and Bindu C S

Division of Civil Engineering, School of Engineering, Cochin University of Science and Technology, Cochin, India

[*]Corresponding author: saranyaullas@gmail.com

ABSTRACT: Inverted pavements can reduce the thickness of granular layers and a thin bituminous layer is provided as the surface course. One of the key functions of the subbase layer is drainage. The current guidelines in India only deal with the strength characteristics of stabilized pavement layers. The permeability characteristics of the subbase course should be given due importance. The suitability of India Road Congress (IRC) recommended gradation for cement-treated subbase (CTSB) layer is considered for the present study. Strength and permeability characteristics were studied by the unconfined compressive strength (UCS) test and using a fabricated permeameter, respectively. The UCS value at 2% cement content was obtained as 2.926 MPa, more than the minimum specified value of 2.25 MPa. The coefficient of permeability was obtained as 0 m/day. Hence the IRC recommended gradation is not suitable for CTSB as a drainage layer.

KEYWORDS: Inverted Pavements, Subbase, Drainage, Permeability

1. INTRODUCTION

The conventional pavement design demands an annual average aggregate requirement of around 450 million tonnes for a high growth scenario in India (Capacity Issues, 2008). Even though aggregates are the least expensive and have the most negligible environmental impacts than other road construction materials, the massive amount significantly affects pavement sustainability (Van Dam). In light of this, government agencies have also restricted quarrying operations, resulting in a severe shortage of non-renewable materials in the construction industry. As a result, these materials have to be transported from remote locations to construction sites across the country, raising the cost of transportation and contributing to greenhouse gas emissions. Options are available for reducing the virgin aggregate consumption in the granular pavement layers and one such method is to stabilize the base/subbase layers with small amounts of cement/bitumen.

The concept of inverted pavements started in South Africa. Indian road congress (IRC) guidelines adopted the stabilized pavement compositions and are included in IRC-37 (2012). The code for stabilization of pavements (IRC-SP-89, 2010) recommends Grading IV of GSB for cement-treated subbase (CTSB) as a drainage layer in inverted pavements. However, very limited studies have taken place in CTSB as a drainage layer. The code also recommends a minimum coefficient of permeability value of 300 m/day for the drainage layer (IRC-37, 2012) and the value is adopted from foreign standards(1000 ft/day or 0.35 cm/s). This paper discusses the suitability of IRC-recommended Grading IV of GSB for CTSB as a drainage layer.

Chapter 8 DOI- 10.1201/9781032657271-8

2. METHODOLOGY

The methodology adopted for the current work is given in the flowchart below.

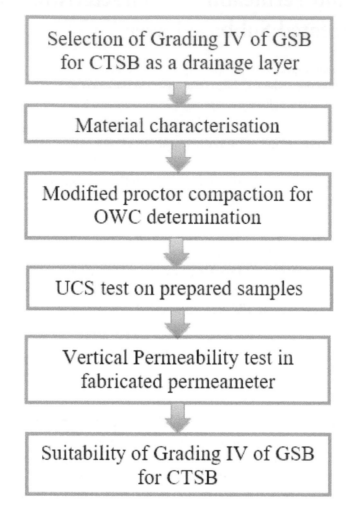

Figure 1. Methodology flow chart

3. EXPERIMENTAL PROGRAM

3.1. Materials

The aggregate materials required for the study were procured from a nearby quarry and the physical properties of the materials satisfied the requirements (Table 1). The ordinary Portland cement 43 grade was used for the study and the physical properties are shown in Table 2.

Table 1. Physical properties of aggregates.

Physical properties of aggregates	Results	Limits (MORTH2013)
Specific gravity (37.5–0.75 mm)	2.751	NA
Abrasion value (%)	35	35 (max)
Impact value (%)	29	40 (max)
Water absorption (%)	0.028	2 (max)

Table 2. Physical properties of cement.

Physical properties of cement	Results	Limits IS: 269 (IS 2015)
Specific gravity	3.125	3–3.25
Consistency (%)	29	25–32
Initial setting time (min)	87	30 (min)
Final setting time (min)	195	600 (max)

3.2. Gradation

The IRC recommended Grading IV of granular subbase was adopted for CTSB and is shown in Table 3.

Table 3. Grading IV of granular subbase (IRC SP 89-2018).

Sieve size (mm)	Percentage passing	
	Lower limit – Upper limit	Adopted gradation
53	100	100
26.5	50–80	65
4.75	15–35	25
0.075	0–5	2.5

3.3. Moisture density relationship

Base/subbase construction has to carry heavy construction traffic over it and requires a compaction density of 98% or higher and recommended test procedure is a modified proctor compaction test. The test was carried out in a cylindrical mold having 150 mm diameter and volume of 2250 cm^3 as per IS 2720 Part 8 (ASTM, 2002; IS 2720-8, 1983). The prepared mixture is compacted with 4.9 kg hammer in 5 layers with 55 blows in each layer, falling from a height of 450 mm (Chhabra et al., 2021; Xuan et al., 2012).

3.3. Unconfined compressive strength (UCS) test

The sample size was 100 mm × 200 mm cylinders based on IS 4332 (1970). The coarse fraction retained on 25 mm sieve was replaced with 25–4.75 mm materials as per IS: 2720 (Part VIII) (IS 2720-8, 1983). The mixture is blended uniformly at optimum water content (OWC), and the mix is then filled in layers using the static compaction method to attain 98% or more density. The test samples were compacted in the molds in five layers using a Marshall hammer (Kasu et al., 2020). The test was conducted at a loading rate of 1.25 mm/min. The UCS test results of cylindrical specimens prepared at OWC (Figure 2a) were reported after curing in moist gunny bags.

3.4. Permeability test

Previous researchers concentrated on the coefficient of vertical permeability of the subbase layer without considering the nominal size of aggregates (Cai et al., 2020; Mousa et al., 2021). The size of the permeameter is usually 8–12 times, the nominal maximum size of the aggregates (ASTM, 2002; Jones and Jones, 1989; Koohmishi and Azarhoosh, 2021). A specially fabricated permeameter of 0.3 m diameter and 0.3 m height was used to determine the coefficient of permeability of the test samples. The test was conducted under fully saturated conditions at constant head. Figure 2(b) shows the vertical permeameter.

Figure 2. (a) Prepared samples and (b) vertical permeameter.

4. RESULTS AND DISCUSSIONS

4.1. Moisture density relationship

In accordance with IRC 37-2018, the recommended cement content for CTSB as a drainage layer is 2–3% (IRC-37, 2012). The compaction test results of the CTSB mixtures are shown in Figure 3. The results indicated that the maximum dry density and OWC increase as the cement content increases. This may be due to the filling of voids with increased cement content.

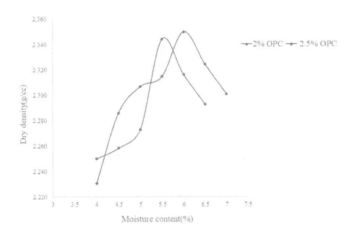

Figure 3. Modified proctor compaction curve.

4.2. UCS test

IRC specifies 1.5 MPa as the minimum strength requirement for CTSBs after 7 days of curing. At the same time, a laboratory value of 1.5 times the minimum specified value (2.25 MPa) is recommended (IRC-37, 2012).

The average strength at 2% cement content was 2.926 MPa, while at 2.5%, the value was 3.668 MPa after 7 days. At 2% cement content, the sample satisfied the strength requirement of 2.25 MPa. The columnar vertical cracking through both ends was identified from the failure pattern (Figure 4).

4.3. Permeability test

The permeability tests revealed that the mix with 2% cement content did not drain and the coefficient of vertical permeability was obtained as 0 m/day. The percentage of filler material plays a vital role in the permeability of the mixture (Huang and Hsien, 2004). The impermeability of the mix may be due to the formation of cementitious gel, which disconnected the pores in the CTSB mix. As a result, IRC recommended grading IV cannot be used for CTSB as a drainage layer. The prepared sample in permeameter and scanning electron microscope (SEM) image of a prepared mixture are shown in Figure 5(a) and (b), respectively.

Figure 4. *Sample at failure.*

Figure 5. *(a) Prepared sample in permeameter and (b) SEM image.*

5. CONCLUSIONS

The suitability of IRC recommended Grading IV of GSB for CTSB as a drainage layer was studied. The strength characteristics of the CTSB mixture satisfied the requirement at 2% cement content. The vertical permeability test at 2% cement content on the prepared CTSB sample showed that the mix was not draining. Thus, the Grading IV of GSB is not suitable for CTSB layer as a drainage layer.

REFERENCES

[1] ASTM, *Standard test method for permeability of granular soils* (constant head) (D2434), 2002.

[2] Cai, X., Wu, K., Huang, W., Yu, J., and Yu, H., "Application of recycled concrete aggregates and crushed bricks on permeable concrete road base," *Road Mat Pav Des,* vol. 22, no.7, pp 1594–1615, 2020.

[3] Capacity Issues, "Indian road construction industry capacity issues, constraints & recommendations," 2008.

[4] Chhabra, R. S., Ransinchung, G. D. R. N., and Islam, S. S., "Performance analysis of cement treated base layer by incorporating reclaimed asphalt pavement material and chemical stabilizer," *Const Build Mat*, vol. 298, p. 123866, 2021.

[5] Huang, and Yang Hsien, *Pavement analysis and design.* Vol. 2. Upper Saddle River, NJ: Pearson Prentice Hall, 2004.

[6] IRC-37, *Guidelines for the design of flexible pavements*, Indian Road Congress, 2012.

[7] IRC-SP-89, *Soil and granular stabilisation*, Indian Road Congress, 2010.

[8] IS 2720-8, *Methods of test for soils*, Part 8: Determination of water content-dry density relation using heavy compaction, Bureau of Indian Standards, 1983.

[9] IS 4332-5, *Methods of test for stabilized soils*, Part 5: Determination of unconfined compressive strength of stabilized soils, Bureau of Indian Standards, 1970.

[10] Jones R. H. and Jones, H. A., "Keynote paper. Granular drainage layers in pavement foundations," in *Unbound aggregates in roads*, Butterworth-Heinemann: Elsevier, pp. 55–69, 1989.

[11] Kasu, S. R., Manupati, K., and Muppireddy, A. R., "Investigations on design and durability characteristics of cement treated reclaimed asphalt for base and subbase layers IIT Pave design," *Const Build Mat*, vol. 252, p 119102, 2020.

[12] Koohmishi, M. and Azarhoosh, A., "Assessment of permeability of granular drainage layer considering particle size and air void distribution," *Const Build Mat,* Vol. 270, p 121373,

[13] Mousa, E., El-Badawy, S., and Azam, A., "Evaluation of reclaimed asphalt pavement as base/subbase material in Egypt," *Transp Geot*, vol. 26, p 100414, 2021.

[14] Van Dam et al., "Towards sustainable pavement systems." A Reference Document.

[15] Xuan, D. X., Houben, L. J. M., Molenaar, A. A. A., and Shui, Z. H., "Mechanical properties of cement-treated aggregate material – A review," *Mat Des*, vol. 33, pp. 496–502, 2012.

9. Feasibility of Using Water Hyacinth as Partial Replacement With Admixture in Glass Powder Incorporated Concrete

Geeja K George[1,*] and Subha Vishnudas[2]

[1]Mar Athanasius College of Engineering, Kothamangalam, Kerala, India

[2]Division of Civil Engineering,
School of Engineering, Cochin University of Science and Technology, Kochi, Kerala, India

[*]Corresponding author: geeja@mace.ac.in

ABSTRACT: The use of reused glass in concrete as a partial replacement to ordinary Portland cement is attracting a lot of interest globally due to the increased discarding cost and environmental issues. Waste glass, when ground to very fine powder shows pozzolanic property which improves the strength of the concrete, though the workability is adversely affected. Chemical and mineral admixtures are commonly used to increase the workability of concrete. This study evaluates the effect of bio-admixture, water hyacinth plant extract, as a replacement for chemical admixture, Conplast SP430. Replacement of 20% cement by glass powder and 20% of Conplast SP430 by water hyacinth extract were found to have higher strength and workability than control mix. Optimum replacement with bio admixture and waste glass in concrete will make the concrete economical and eco-friendly without affecting the mechanical properties.

KEYWORDS: Glass Powder (G), Water Hyacinth Extract (H), Workability, Optimum Replacement, Eco-Friendly

1. INTRODUCTION

Alternative binders to portland cement have been suggested to reduce greenhouse gas emissions as blended cements (Rashad, 2015). Mechanical properties of concrete were studied with different types of recycled glass utilized as partial replacements for cement at varying concentrations, and it was found that compressive strength could increase at an optimum replacement of 20% cement by glass powder (Islam et al., 2017; Khan et al., 2020). Waste glass, when ground to very fine powder shows pozzolanic properties which improves the strength of the concrete (Shi et al., 2005). But the workability of waste glass replaced concrete will decrease as the amount of glass powder increases (Safarizki et al., 2020). Many researchers have attempted the replacement of cement or fine aggregate with glass powder in concrete in different proportions to study the mechanical behavior of sustainable concrete to promote the reuse of glass materials (Andreola et al., 2016; Bostanci et al., 2016; Du and Tan, 2017; Patel and Dalal, 2017). Nowadays, admixtures are being used more frequently in the concrete industry, particularly bio-admixtures (Islam et al., 2017). Water resources and aquatic life are severely harmed by water hyacinth (Abdel Sabour, 2010). Research on effective methods for controlling water hyacinths is still going on. This study evaluates mechanical properties due to the effect of bio-admixture, water hyacinth plant extract as a partial replacement for the admixture, Conplast SP430 in glass powder incorporated—concrete. All of that is carried out to produce concrete with a reduced environmental impact that is related to the production of cement, as well as alternate methods for sustainable management and potential water hyacinth plant utilization (Okwadha and Makomele, 2018).

Chapter 9 DOI- 10.1201/9781032657271-9

1.1. Aim and scope of the study

As concrete is the most widely used construction material in the world, the disposal of waste glass as a partial replacement for cement is considered to make it eco-friendly and less expensive. By adding admixtures, the workability of glass powder added concrete can be improved. A bioadmixture, water hyacinth makes the construction industry eco-friendlier and less expensive. Investigating the efficacy of using water hyacinth extract as a bio admixture for replacing Conplast SP430 in glass powder-incorporated concrete is the goal of this research.

2. MATERIALS AND METHODS

2.1. Materials

The commercially available glass powder is collected. Material characterization of ordinary Portland cement of 53 grade was done according to relevant Indian standards (IS 12269) (2013) and is used for the study. Coarse aggregates with a maximum size 20 mm were used. Material characterization and the grain size analysis of the aggregate sample were conducted according to the procedure conforming to the IS 383 (1970). Manufactured sand is used as fine aggregate and its specific gravity and water absorption are found to be 2.75% and 1.9%, respectively. The sand used falls in Zone II according to the standards. Water hyacinth plant collected, is powdered after cleaning and drying. 500 g of the powder is wetted with 1 L of clean potable water and soaked in 30 mL of ethanol for 24 hours. The extract is collected by filtering. The components of water hyacinth extract are identified by gas chromatography–mass spectrometer (GC–MS) analysis (Okwadha and Makomele, 2018). Conplast SP 430 significantly improves the workability of the concrete without increasing water demand (Elinwa, and Maidawa, 2018). Potable water is used for concreting and the w/c ratio is adopted as 0.45.

2.2. Methodology

M40 mix is prepared in different proportions with glass powder and water hyacinth extract and analyzed the mechanical properties of concrete.

2.2.1. Mix Design Samples of Glass Powder, Conplast Sp 430, and Water Hyacinth Incorporated Concrete

Concrete mix design is prepared as per IS: 10262 (2019). Mix design samples are coded as GH- x%, representing, partial replacement with glass powder (G) and water-hyacinth (H) by x% individually for cement and admixture, respectively, with an incremental replacement of 5%. Mix designs for different proportions of glass powder, Conplast SP430 and water hyacinth incorporated concrete are shown in Table 1. GH-0% represents the control concrete. Slump test on fresh concrete samples were conducted as per IS: 1199 (2018). Compressive strength test on concrete cubes was conducted as per IS: 516 (1959) for after 7 and 28 days of curing. Split tensile strength on concrete was conducted as per IS: 5816 (1999) after 7 and 28 days of curing. Water absorption test on concrete was conducted as per IS: 1124 (1974).

Table 1. Mix design of glass powder, conplast SP430, and water hyacinth-incorporated concrete in various proportions.

Sample	Cement (kg)	Coarse aggregate (kg)	Fine aggregate (kg)	Glass powder (kg)	Water (l)	Conplast SP 430 (ml)	Water hyacinth extract (ml)
GH-0%	11	39.25	19.6	0	3.7	60	0
GH-10%	9.9	39.25	19.6	1.1	3.7	54	6
GH-15%	9.35	39.25	19.6	1.65	3.7	51	9
GH-20%	8.8	39.25	19.6	2.2	3.7	48	12
GH-25%	8.25	39.25	19.6	2.75	3.7	45	15

3. RESULTS AND DISCUSSIONS

3.1. Workability—slump test

The workability of concrete is measured from slump test and Figure 1 shows the slump value of various mixes on fresh concrete. Since the fine glass powder absorbs excess water, the slump value of the fresh concrete has decreased as the % replacement of glass powder increased relative to the control concrete (Safarizki et al., 2020). The trend for workability was unchanged as a result of the bioadmixture gradually replacing Conplast SP430; however, an improvement in slump can be observed in contrary to the replacement solely with glass powder. In addition, the percentage reduction in slump with reference to the control mix for GH-20% is (optimum mix based on strength) less than that in G-20% with reference to the control mix. This indicates that the water hyacinth extract decreases the rate of hydration and hardening, enabling excellent flow ability (Okwadha and Makomele, 2018).

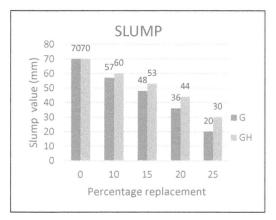

Figure 1. *Slump values as a function of different percentage replacements.*

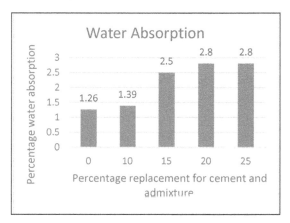

Figure 2. *Water absorption percentage of concrete specimens.*

3.2. Water absorption test on concrete

Water absorption is defined as the amount of water absorbed by a material and the water absorption percentage is given in Figure 2. Water absorbing property of the mix samples increases as the percentage replacement for cement and admixture increases (Figure 2). The durability of the concrete may be affected by a higher water absorption value when the concrete becomes porous. Consequently, more research may be carry out on the admixture's water absorption and to include a water reducer. However, as the quantity of water hyacinth extract increased, the incremental rate dropped, which is consistent with the findings of various researchers (Okwadha and Makomele, 2018).

3.3. Compressive strength and split tensile strength

The compressive test was conducted for after 7 and 28 days of curing and the result is shown in Figure 3. Compressive strengths with the replacement of bio admixture and glass powder to control-concrete were showing an increasing trend till 20% replacement both in 7 days and 28 days of curing. The results agree with the earlier investigation by Elaqra et al. (2019). Waste glass, when ground to very fine powder shows pozzolanic property which improves the strength of the concrete (Shi et al., 2005). Fatty acids in the extract may be the contributing factors why compressive strength increases as water hyacinth extract concentration increases. After 20% replacement, compression strength decreases due to high fluidity causing segregation and bleeding of the matrix, could have decelerated hydration and strength gain creating a porous permeable matrix (Islam et al., 2017; Okwadha and Makomele, 2018). The test for split tensile strength was conducted on concrete cylinders after 7 and 28 days of curing. The results obtained for different mixes are illustrated in Figure 4. The trend for split tensile strength with different percentage replacement of cement and admixture shows similar to the trend of compressive strength.

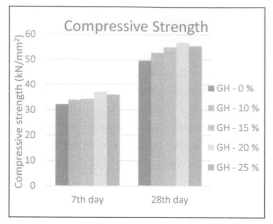

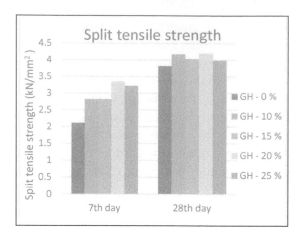

Figure 3. Compressive strength of concrete samples for different ages.

Figure 4. Split tensile strength of concrete specimens at different ages.

4.   CONCLUSIONS

The conclusions drawn from this study are summarized below.

- The optimum replacement of water hyacinth was found to be 20%.
- The optimum replacement of glass powder was found to be 20%.
- Compressive and tensile strengths at the optimum replacement are greater than the normal concrete.
- With the addition of glass powder, the strength of concrete increases, but the workability of concrete decreases. Addition of water hyacinth will compensate the same.

The study suggests that using a water reducer with the material the water absorption is increased. It is also recommended that more research be done on the blend material's capacity to absorb water. In general, considering the performance with replaced material, glass can reduce the cement production, results in the reduction of CO_2 emission, and saves the environment significantly by reducing greenhouse gas (Islam et al., 2017). The findings show that water hyacinth weed can be used economically as a co-superplasticizer to substitute a chemical admixture, making the building sector more environmentally friendly and less expensive.

REFERENCES

[1]   Abdel Sabour, M. F., "Water hyacinth: Available and renewable resource," *Electr J Environ Agri Food Chem*, pp. 1743–1759, 2010.
[2]   Andreola, F., Barbieri, L., Lancellotti, I., Leonelli, C., and Manfredini, T., "Recycling of industrial wastes in ceramic manufacturing: State of art and glass case studies," *Ceramics Int*, vol. 42, no. 12. pp. 13333–13338, 2016.
[3]   Bostanci, ,Limbachiya, M., and Kew, H., "Portland-composite and composite cement concretes made with coarse recycled and recycled glass sand aggregates: Engineering and durability properties," *Constr Build Mat*, vol. 128, pp. 324–340, 2016.
[4]   Du, H., and Tan, ,"Properties of high-volume glass powder concrete," *Cement Conc Comp*, vol. 75, pp. 22–29, 2017.

[5] Elaqra, H. A., Haloub, M. A. A., Rustom, R. N., "Effect of new mixing method of glass powder as cement replacement on mechanical behavior of concrete," *Constr Build Mater*, vol. 203, pp. 75–82, 2019.

[6] Elinwa, , and Maidawa, S., "Mechanical strength of admixed conplast SP 430 cement paste and concrete", 2018. Available at SSRN 3174640.

[7] IS 10262 – (2019), *Concrete Mix Proportioning Guidelines*, New Delhi: Bureau of Indian Standards, 2019.

[8] IS 1124 – (1974), *Method of Test for Determination of Water Absorption, Apparent Specific Gravity*, New Delhi: Bureau of Indian Standards, (Reaffirmed 2003), 1974.

[9] IS 1199–(2018), *Fresh Concrete – Methods of Sampling, Testing and Analysis. Part 5 Making and Curing of Test Specimen*, New Delhi: Bureau of Indian Standards, 2018.

[10] IS 12269 – (2013), *53 Grade Ordinary Portland Cement*, New Delhi: Bureau of Indian Standards, 2013.

[11] IS 383 – (1970), *Specification for Coarse and Fine Aggregates from Natural Sources for Concrete*, New Delhi: Bureau of Indian Standards, 1970.

[12] IS 516 – (1959), *Method of Tests for Strength of Concrete*, New Delhi: Bureau of Indian Standards, 1959.

[13] IS 5816 – (1999), *Splitting Tensile Strength of Concrete Method of Test*, New Delhi: Bureau of Indian Standards, 1999.

[14] Islam, G. M. S., Rahman, M. H., Kazi, N., "Waste glass powder as partial replacement of cement for sustainable concrete practice," *Int J Sustain Built Environ*, vol. 6, no. 1, pp. 37–44, 2017.

[15] Khan, M. N. N., Saha, A. K., Sarker, P. K., (2020) "Reuse of waste glass as a supplementary binder and aggregate for sustainable cement-based construction materials: a review," *J Build Eng*, vol. 28, 2020.

[16] Okwadha, G. D. O., and Makomele, D. M., "Evaluation of water hyacinth extract as an admixture in concrete production," *J Build Eng*, vol. 16, pp. 129–133, 2018.

[17] Parel, H. G., and Dalal, S. P., "An experimental investigation on physical and mechanical properties of concrete with the replacement of fine aggregate by poly vinyl chloride and glass waste," *Procedia Eng*, vol. 173, pp. 1666–1671, 2017.

[18] Rashad, A. M., "A brief on high-volume Class F fly ash as cement replacement—a guide for Civil Engineer," *Int J Sustain Built Environ*, vol. 4, no. 2, pp. 278–306, 2015.

[19] Safarizki, H. A., Gunawan, L. I., and Marwahyudi, "Effectiveness of glass powder as a partial replacement of sand in concrete mixtures," *J Phys Conf Series*, vol. 1625, no. 1, 2020.

[20] Shi, C., Wu, Y., Riefler, C., and Wang, H., "Characteristics and pozzolanic reactivity of glass powders," *Cement Conc Res*, vol. 35, no. 5, pp. 987–993, 2005.

Theme: Environmental Engineering

10. Life Cycle Assessment of Edible Water Blobs Using Openlca

Catherine Chakkalakkal, Benita Nishil, Vishnudatha V and Ambika S*

Department of Civil Engineering, Indian Institute of Technology Hyderabad, Sangareddy, Telangana, India

*Corresponding author: ambika@ce.iith.ac.in

ABSTRACT: The use of plastic water bottles is ineluctable. The plastic bottles that are thrown away after its use without proper disposal and recycling have invaded our ecosystem. Various research are done on biodegradable packaging materials. Edible water blob is an innovative technology evolved from the thought of biodegradable packaging. This study aims to understand the different stages in the production process of edible water blob which contributes to greater environmental impact. In this study, life cycle assessment (LCA) of edible water blobs made from seaweed extract and lignocellulosic biomass is considered. LCA is performed according to the ISO 14040/44 standard methodology. The LCA models are developed with cradle-to-gate concept using OpenLCA software and AGRIBALYSE database to quantify the life cycle impacts of 50 numbers of edible water blobs. Those results are used to interpret various environmental impact categories of which global warming potential, land use, and water consumption are interpreted in detail. This is the first study on LCA of edible water blobs done based on secondary data collected from various literature. This study aims to analyze the impact of the product before its commercialization on a large scale. The assessment shows that calcium lactate production impacts ecosystem and resource availability more than sodium alginate while vice versa for health impacts is observed. The study showed that sodium alginate production was prone to cause more health impact and fossil fuel scarcity while calcium lactate impacts the ecosystem.

Keywords: Edible Water Blobs, OpenLCA, Life Cycle Assessment, Sodium Alginate, Calcium Lactate

1. INTRODUCTION

Edible water blob is an emerging technology to replace plastic water bottles. The edible sachets are made from seaweed extract and lignocellulosic biomass. The production of plastics since the 1950s has shown an exponential increase such that by 2050, there will be more plastics than fish in the ocean. The global plastic production in 2019 reached 368 million metric tons (UNEP). The single use of plastic bottles among them is an integral part of our lives. The bottles are made of polyethylene terephthalate, and it takes nearly 450 years to degrade completely (WWF). Thus, edible water blobs which are made of biodegradable material, that takes 6 weeks to degrade, can revive our planet to an extent (OOHOWATER, 2019). Edible water blobs named Ooho were first introduced as a trial in the London marathon by an innovative sustainable packaging company named Skipping Rocks Lab based in London to serve the participants with energy booster drinks in a bid to reduce plastic waste (*Independent*). Although the product is supposed to be environmentally sustainable, before the commercialization of the product on a large scale, its life cycle assessment (LCA) would help to get an insight on the environmental impacts it poses and the environmental hotspots in the lifecycle of the product. In this study, water blobs are made from sodium alginate, which is a seaweed extract and calcium lactate which is obtained from sugarcane bagasse. This study aims to perform the LCA of edible water blobs using OpenLCA software, analyze the inputs and output of materials during the life cycle of edible water blobs, and examine the various environmental impacts of the product.

2. MATERIALS AND METHODS

Life cycle assessment or LCA study was performed according to the standard methodology described in the ISO 14044 by the International Organization of Standardization (ISO). This study was done as a cradle-to-gate LCA of edible water blobs. The LCA model was made using OpenLCA version 1.10.3 and AGRIBALYSE database. In this study, 50 numbers of edible water blobs were taken as the functional unit, which together amounts to 1 L of drinking water.

3. PROCESS DESCRIPTION

The raw materials used in the production of edible water blobs are calcium lactate and sodium alginate.

3.1. Production of calcium lactate

Calcium lactate is a white crystalline salt with the chemical formula $C_6H_{10}CaO_6$ made from lactic acid and calcium hydroxide (PubChem, 2005a). In this study, lactic acid production from the fermentation of sugar cane bagasse which is a waste product obtained after the extraction of juice from sugar cane is considered (Shastri et al., 2021). The inputs for 1 tonne of sugarcane from 0.0256 ha of farmland (Shukla and Kumar, 2019). The farmland is irrigated with 342.8 m^3 of water (Shukla and Kumar, 2019). The agricultural machinery used in plowing and threshing activity operates with diesel fuel. About 0.64 t of bio compost, 7.68 kg of N, 5.12 kg of K_2O, and 2.56 kg of P_2O_5 fertilizers are applied in the field (Shukla and Kumar, 2019). The pesticide use of atrazine at the rate of 2 kg ai per ha is considered (Shukla et al., 2017). The fertilizers and pesticides are transported through a distance of 50 km from the regional production plant. This study considers the practice of post-harvest burning which assumes about 25–30% of plant mass contributes to green waste (Mashokoa et al., 2010). The sugarcane is harvested manually and transported to the sugar mill. The sugar bagasse from the sugar mill after extracting the juice is used in the production of lactic acid in the biorefinery. About 1000 kg of sugarcane leads to the formation of 280 kg of sugarcane bagasse (Mashokoa et al., 2010). The lactic acid production in the biorefinery involves a series of steps such as size reduction of bagasse, pre-treatment, hydrolysis, fermentation, acidification, and filtration (Shastri et al., 2021). The processes in biorefinery require the use of various chemicals such as sodium hydroxide and sulfuric acid. The fermentation process requires calcium carbonate to neutralize the low pH conditions which have an inhibitory effect on microbial activity and yeast extract is used as fermentation activator (Shastri et al., 2021). The filtration process removes the impurities present in the lactic acid produced. The lactic acid from biorefinery is reacted with food-grade calcium hydroxide to form calcium lactate (PubChem, 2005a). In the salt processing unit, it is dried and crushed to form calcium lactate powder. The calcium lactate is packed and transported. The calcium lactate is packed and transported over 50 km.

3.2. Production of sodium alginate

Sodium alginate ($C_6H_9NaO_7$) (PubChem, 2005a) is a naturally occurring anionic polymer obtained from brown seaweeds (McHugh, 2003). The brown seaweed, kelp, which belongs to the Laminariacea family and good source of alginates, was taken for the study. The plantlets were grown from the spores collected from wild seaweed and cultivated in concrete ponds in a nursery. The nursery was in a closed building which was taken as an agricultural shed. Materials such as mineral fertilizers, spargers for bubbling, fluorescent lamps, booster, and circulation pumps were used in the cultivation of plants. Pumped and filtered seawater was used for the plantlet cultivation and treated with UV radiation to inhibit any microbial growth (Langlois et al., 2012). The plantlets from the nursery were taken offshore and tied to anchored floating lines. Algae were attached to culture ropes which were connected to structural ropes. This was stabilized by concrete blocks at regular spacings (Langlois et al., 2012). During growth, the algae take up carbon dioxide for photosynthesis and release oxygen. The nutrients such as nitrates and phosphates are taken up from the ocean environment. In this study, oxygen and water released in respiration are taken into account. The dry weight to fresh weight ratio is taken as 0.1 (Duarte et al., 2017) for calculations. The harvesting and transport were carried out by boat. The amount of fuel required for this purpose was considered for calculations. The harvested biomass was washed and crushed. Bleaching was done with alcohol. The mass-to-volume ratio for alcohol treatment was taken as 1:50 (Shukla et al., 2017). The soluble fraction from alcoholic pre-treatment was extracted and dewatered using 30% HCl solution. Alginate was then solubilized using 1.5% sodium carbonate solution. After dewatering, cellulose powder was used to extract the residues. The acid was then precipitated using 30% HCl solution (Boonstra, 2015). It was again dewatered, and the temperature was brought down to 4°C. Sodium carbonate was added to convert alginic acid to sodium alginate. It was then dried with a dryer. The amount of sodium alginate obtained per kg of dry matter of algae is 0.42 kg (Boonstra, 2015). The alginate powder is then packed and transported.

3.3. Production of edible water blobs

One edible water blob is assumed to contain approximately 20 ml of drinking water. For 1 l of water, 50 water blobs are required. The following steps are followed to make 50 water blobs (Mistry and Kharate, 2020): (i) calcium lactate solution is made by dissolving 20 g calcium lactate in 4 l of water by stirring well. And sodium alginate solution is made by dissolving 4 g sodium alginate in 1 l of water with 2-minute mixing using an electric blender. Then, a scoop of sodium alginate solution (20 ml) is taken and carefully immersed into calcium lactate solution. The calcium lactate solution is kept unstirred until it forms spherical blobs of water. The water blob is carefully taken from the solution and rinsed in a bowl of pure water.

4. RESULTS AND DISCUSSION

4.1. Life cycle impact assessment

Table 1 shows the impacts of production of 50 numbers of edible water blobs as per the ReCiPe (H) midpoint assessment method. It gives 18 impact categories.

Table 1. Life cycle impacts for production of 50 numbers of edible water blobs.

Impact category	Value
Terrestrial ecotoxicity	1.86059kg 1,4-DCB
Fossil resource scarcity	0.5023 kg oil eq.
Marine ecotoxicity	0.0237 kg 1,4-DCB
Mineral resource scarcity	0.00361kg Cu eq.
Terrestrial acidification	0.0028 kg SO_2 eq.
Global warming	1.1860 kg CO_2 eq.
Ozone formation, terrestrial ecosystems	0.00262kg NO_x eq.
Fine particulate matter formation	0.00145 kg $PM_{2.5}$ eq.
Freshwater ecotoxicity	0.01685 kg 1,4-DCB
Ionizing radiation	0.06771 kBq Co-60 eq.
Water consumption	0.02645 m^3
Ozone formation, human health	0.0024 kg NO_x eq.
Human carcinogenic toxicity	0.02704 kg 1,4-DCB

Stratospheric ozone depletion	2.697E-6kg CFC11 eq.
Freshwater eutrophication	0.00037 kg P eq.
Land use	0.15522 m²a crop eq.
Human non-carcinogenic toxicity	0.75145 kg 1,4-DCB
Marine eutrophication	9.44E-5 kg N eq.

The process-wise breakdown of impacts of sodium alginate production, calcium lactate production, tap water production, and electricity in blending with the percentage contribution of each process in the production of edible water blobs is shown in Figure 1. In this study, major impact categories such as global warming potential, land use, and water consumption are considered in detail as shown in Figure 2. The global warming potential is 1.186 kg CO_2 eq. 59.72% is contributed by sodium alginate production and 39.94% by calcium lactate production. An in-depth analysis of stage wise impact of the process has shown that 48.44% is caused by ethanol production for its use in alcoholic pre-treatment of algae (bleaching). In calcium lactate production, the sugarcane cultivation stage has contributed to 33.1% of the impact. The land use is 0.15525 m² a crop eq. in which 90.62 % is contributed by calcium lactate production and 9.37% by sodium alginate production. In sodium alginate production, sugarcane cultivation causes 82% of the impact. In sodium alginate production, ethanol production causes 4% of the impact. The water consumption is 0.02645 m³, 32.53% of water is consumed in sodium alginate production, 41.98% in calcium lactate production, 22.73% in tap water production, and 2.77% in electricity production for blending. The impact analysis shows that the processes contributing to health impacts categories are found to be those involved in the production of sodium alginate. Calcium lactate production is contributing more to impacts associated with damage to the ecosystem and resource availability while sodium alginate production also causes fossil resource scarcity.

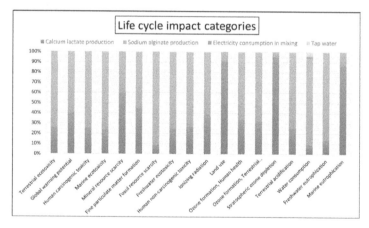

Figure 1. *Life cycle impact categories for 50 numbers of edible water blob production.*

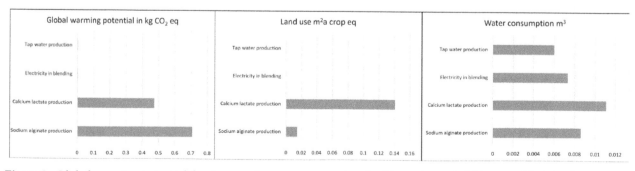

Figure 2. *Global warming potential, land use, and water consumption for 50 numbers of edible water blob production.*

5. CONCLUSION

The LCA of edible water blobs was performed with a cradle-to-gate approach and inputs and outputs of the processes involved were analyzed. The life cycles of calcium lactate and sodium alginate in the production of water blobs were also analyzed. The impacts on health, ecosystem, and resource availability were calculated using OpenLCA software. The output showed that sodium alginate production was prone to cause more health impact and fossil fuel scarcity while calcium lactate impacts the ecosystem. Though the edible water blob is environmentally safe due to its biodegradability, this study has proven the hidden impacts due to the raw materials. To further reduce the impact on health and fossil fuel, sustainable agricultural practices, different alternatives for production of calcium lactate or different brown algae can be tested and adopted.

6. FUTURE SCOPE

This study considers edible water blob preparation from sodium alginate and calcium lactate, various other alternatives such as corn stover for production of calcium lactate, food grade calcium hydroxide, different brown algae for sodium alginate extraction, and a comparative LCA study can be done. In depth analysis of the biorefinery stage in the production of sodium alginate and calcium lactate can be done to suggest alternatives to reduce the impacts. In this study, 5:1 mixing ratio of calcium lactate and sodium alginate is used for the preparation of water blobs, other ratios are to be tested with. Research is going on to produce a packaging of more than 10 blobs together to make it more user-friendly.

REFERENCES

[1] Boonstra, A. S., "The macroalgae-based biorefinery – a comprehensive review and a prospective study of future macroalgae-based biorefinery systems," Master's thesis. Utrecht University, Utrecht, Netherlands, 2015.

[2] Duarte, C. M., Wu, J., Xiao, X., Bruhn, A., and Krause-Jensen, D., "Can seaweed farming play a role in climate change mitigation and adaptation?," *Front Mar Sci*, vol. 4, 2017.

[3] *Independent*. https://www.independent.co.uk/life-style/london-marathon-edible-water-bottle-capsules-a8889796.html

[4] Langlois, J., Sassi, J. F., Jard, G., Steyer, J. P., Delgenes, J. P., and Hélias, A., "Life cycle assessment of biomethane from offshore-cultivated seaweed," 2012.

[5] Mashokoa, L., Mbohwab, C., and Thomas, V. M., "LCA of the South African sugar industry," *J Environ Plan Manag*, vol. 53, no. 6, pp. 793–807, 2010.

[6] McHugh, J. D., "A guide to the seaweed industry," FAO Fisheries Technical Paper, 441, Food and Agriculture Organization of The United Nations, Rome, 2003.

[7] Mistry, U., and Kharate, S. P., "A study of edible water balls in Indian market context," Corporate Social Responsibility & Sustainable zevelopment, Pune, 10–12 December, 2020.

[8] OOHOWATER, 2019. http://www.oohowater.com/ [accessed 15 November 2021].

[9] PubChem, "Calcium lactate," National Library of Medicine, 2005a. https://pubchem.ncbi.nlm.nih.gov/compound/calcium-lactate [accessed 1 November 2021].

[10] PubChem, "Sodium alginate," National Library of Medicine, 2005b, https://pubchem.ncbi.nlm.nih.gov/compound/sodium-alginate [accessed 1 November 2021].

[11] Shastri, Y., Munagala, M., Nalawade, K., Konde, K., and Patil, S., "Life cycle and economic assessment of sugarcane bagasse valorization to lactic acid," Waste Manag, vol. 126, pp. 52–64, 2021.

[12] Shukla, A., and Kumar, S. Y., "Life cycle analysis of sugar production process in central India," *Int J Sci Technol Res*, vol. 8, no. 11, 2019.

[13] Shukla, S. K., Sharma, Lalan, Awasthi, S. K., and Pathak, A. D., "Sugarcane in India: Package of practices for different agro-climatic zones," ICAR-All India Coordinated Research Project on Sugarcane. pp. 1–64, 2017.

[14] UNEP. https://www.unep.org/interactive/beat-plastic-pollution/

[15] WWF. https://www.wwf.org.au/news/blogs/the-lifecycle-of-plastics#gs.h03hu9

[16] Yuan, S., Duan, Z., Lu, L., Ma, X., and Wang, S., "Optimization of decolorization process in agar production from *Gracilaria lemaneiformis* and evaluation of antioxidant activities of the extract rich in natural pigments," *Biotech*, vol. 8, no. 1.

11. Performance Assessment of Constructed Wetland in Treatment of Wastewater

Akshay Ranjith[1]*, Alisha P M[3], Binsy P T[3], Catherine Chakkalakkal[2] and Sumi S[3]

*[1]Department of Civil Engineering, Indian Institute of Science, Bangalore, Karnataka

[2]Department of Civil Engineering, Indian Institute of Technology Hyderabad, Sangareddy, Telangana, India

[3]Department of Civil Engineering, College of Engineering Trivandrum, Thiruvananthapuram, Kerala, India

Corresponding author: rakshay@iisc.ac.in, catherine58cet@gmail.com, pmalisha245@gmail.com, binsychandran1234@gmail.com, sumis@cet.ac.in

ABSTRACT: Septic tank-soak pit system is the most adopted system for sewage treatment for small communities in our country. The wastewater treatment facility of the Ladies Hostel (LH) of the College of Engineering Trivandrum (CET) also makes use of a similar system. However, the increase in the number of inmates in recent years has resulted in higher influent loading. As a sustainable, energy-efficient, and economical solution for this, this study aims to solve the environmental problems faced by the LH of CET by effectively using the constructed wetland system in the hostel premises. This report examines the performance of constructed wetland in sewage treatment. The study also looks into the individual contribution of the filter media and the plant toward contaminant removal using secondary data. A phytoremediation model will be developed using the results from the performance tests.

KEYWORDS: Canna indica, Efficiency, Phytoremediation, Sewage, Wetland

1. INTRODUCTION

1.1. Constructed wetlands

Natural wetland systems are often referred to as the "earth's kidneys" due to their ability to filter pollutants from water (Joseph, 2018). Constructed wetlands (CWs) are engineered systems that emulate natural wetlands by utilizing plants, soils, and microorganisms to remove contaminants. CWs are cost-effective, energy-efficient, and sustainable engineering solutions. Different types of CWs, including horizontal subsurface flow, vertical subsurface flow, and hybrid subsurface flow, exist, with vertical flow CW being the most efficient (Ramaprasad and Philip, 2016; Vymazal, 2010). Previous studies have reported the removal of contaminants in CWs operating in cold climates using free water surface, subsurface flow, and hybrid wetland systems (Herath and Vithanage, 2015; Wang et al., 2017). Removal efficiencies of parameters such as BOD_5, COD, nitrogen, and phosphorous were observed to decrease in winter, with higher removal of nitrogen in summer (Herath and Vithanage, 2015; Wang et al., 2017). Plant uptake was found to be more effective in removing trace elements during the summer season (Jacome, 2016; Ramaprasad, C. and Philip, 2016; Wang et al., 2017). The radial oxygen loss (ROL) process occurring in the root zone of macrophytes contributes to pollutant removal by creating an aerobic micro-environment (Cui et al., 2010; Sandoval et al., 2019; Zhang et al., 2007). Bioremediation processes, principles, and mechanisms have been extensively studied (Vidali, 2001). Among various plant species analyzed, Scirpus validus and water hyacinth were found to be effective in phytoremediation of phosphorus and phytoextraction of manganese, respectively (Sandoval et al., 2019). Ornamental plants, such as Canna spp., Iris spp., Heliconia spp., and Zantedeschia spp., are commonly used in CWs for aesthetic purposes. *Canna indica* has shown high purification efficiencies in the treatment of sewage water and greywater (Joseph, 2018). It has also been identified as a suitable option for high-strength wastewater treatment under tropical conditions (Haritash et al., 2015). However, different observations have been made regarding the performance of *C. indica* compared to other plant species. *C. indica*, with its fine fibrous roots, provides a larger surface area for the attachment of nitrogen-fixing bacteria, leading to increased nitrogen and phosphorous removal (Sugyana and Sebastian, 2017). The plant has also demonstrated potential for phytoextraction of heavy metals from contaminated soils.

1.2. Need for the study

The sewage treatment system at CET's Ladies Hostel (LH), initially designed for around 500 students, is facing overload issues due to the increased number of inmates, which now stands at 750. This study aims to evaluate the effectiveness of the existing CW when treating septic tank effluent to address this problem. The approach involves using C. indica plants for phytoremediation. By integrating the septic tank with the CW, it is expected to alleviate the overload on the septic tank and effectively address the current problems in the Ladies Hostel. The paper also focuses on analyzing the individual efficiencies of the plant and filter media, ultimately developing a phytoremediation model.

1.3. Existing CW

The VF CW, constructed in 2018–2019, measures 13 m × 11 m × 1.2 m. It operates with subsurface vertical flow, filled with gravel, and has a capacity of 30,000 l/day with intermittent loading. This CW serves as both a growth bed and a filter bed, supporting vegetation. For more information, please refer to (Joseph, 2018) for detailed construction parameters.

2. METHODOLOGY

The study was conducted in five stages and the methodology adopted for the study is as follows: (i) characterization of influent and effluent of existing CW, (ii) design of laboratory scale CW, (iii) development of laboratory scale CW, (iv) sampling and analysis of treated wastewater, and (v) modelling CW.

2.1. Characterization of wastewater

The treated effluent from the existing CW was regularly analyzed and monitored on a monthly basis. Influent and effluent samples were analyzed for various parameters including BOD, COD, alkalinity, acidity, TS, pH, turbidity, and chlorides following APHA standards.

2.2. Design procedure of laboratory scale CW

The design procedure of CW unit was adopted from Environment Protection Agency (EPA) standards (2000 and 2008). Computing procedure for the design of a vertical subsurface CW is as follows: (i) determine the media type, vegetation, and depth of bed to be used, (ii) assume the porosity and effective hydraulic conductivity of the media to be used, (iii) determine the influent BOD concentration and water flow rate, (iv) assume the effluent BOD concentrations, (v) determine the bed surface area, and (vi) determine bed length and width for an assumed length to width ratio. Using Darcy's law, the flow that can pass through the bed surface was determined. The flow is possible if the resulting flow rate is less than the actual design flow. The design based on first-order plug flow kinetics was used in this study. This method is based on design criteria using flow rate and organic loading rate.

2.3. Sampling and analysis of treated wastewater

A monthly sampling and analysis using synthetic wastewater as the influent was proposed to be conducted for the parameters BOD_5, COD, nitrates, phosphates, pH, and TDS. But, due to the closing of colleges as per the order of the Government of Kerala in the wake of COVID-19 pandemic on 10 March 2020, the further procedures of the work had to be modified. For the calculation of the efficiency of the filter medium and plant medium separately, secondary data on the performance of the lab scale CW with *C. indica* had to be depended upon.

2.4. Modeling CW

Due to non-availability of sufficient data from the experimental studies, modeling study was carried out based on secondary data based on a lab-scale study set up on the campus of Delhi Technological University (Haritash, et al., 2015). The wastewater collected from a collection in Delhi Technological University campus, which receives wastewater from nearby residential blocks and partially from a university hostel was used as the influent. The plant used for phytoremediation was *C. indica*. The modeling of the lab scale CW has been carried out using SPSS software and charts for direct analyses were prepared using MS Excel.

3. RESULTS AND DISCUSSION

3.1. Basic characterization

The initial complete study of the influent and effluent of the CW was thoroughly done with the investigation of all possible parameters with respect to the treatment of wastewater. The results of the basic characterization of existing CW are tabulated in Table 1.

3.2. Design of laboratory scale CW

For the design of laboratory scale model of CW, Q was assumed as 30 l/day, d = 0.6 m, C_o = 200 mg/l, and C_c = 20 mg/l. By experiment, n = 0.27 and from the existing CW dimensions, L/B ratio = 13/11. From equation (1), area of surface bed A_s = 0.2156 m², L = 0.51 m, and B = 0.43 m. From equation (2), HRT was obtained, t = 1.1842 days. From equation (3), effluent BOD C_e = 19.24 (<20 mg/l, assumed value). From equation (4), average daily flow Q = K_s A S = 0.000 m³/day (<0.03 m³/day). Here, Darcy's law, equation (4), was not used, as the bed slope in the case of the lab scale CW is negligible. From equation (5), area loading rate, ALR = 27.83 g/m²/day > 6 g/m²/day.

Table 1. Characterization results.

Parameters	Influent (sewage)	Effluent	Drinking water standards[a]	Effluent standards[b]
pH	8.05	7.13	6.5-8.5	6.5-9
Temperature (°C)	29	29	--	--
Conductivity (µS/cm)	1160.0	302.4	--	--
Turbidity (NTU)	8	4	1	2-5
Iron (µg/L)	11.79	16.38	300	3000
Dissolved oxygen (mg/l)	0	5.1	--	--
BOD_5 (mg/l)	59.6, 320	0	0	30
Sulfates (mg/l)	3.133	0.34	200	--

Phosphates (µg/l)	5.268	3.103	--	--
Alkalinity (mg/l of CaCO₃)	1135	140	200	--
Acidity (mg/l of CaCO₃)	260	105	--	--
Chlorides (mg/l)	67.498	44.99	250	--
Total dissolved solids (mg/l)	608	157.3	500	2000
Nitrites (mg/l)	3.424	0.339	0	--

[a]IS 10500: 2012 (drinking water standards).
[b]Standards and guidelines KSPCB 2016.

Table.2. Characteristics of laboratory scale model CW.

Flow pattern	Subsurface vertical flow
Bed dimension	0.5 m x 0.43 m
Filter Media	M sand and aggregate
Bed depth	0.6m
Surface area	0.225 m2
Average flow	30 L/day
Type of waste water	Synthetic black water
Type of loading	Intermittent

The characteristics of the laboratory scale model CW are shown in Table 2. An intermittent type of loading was provided in the laboratory scale CW with an average flow of 30 l/day. The hydraulic retention time was obtained as 1.2 days. A subsurface vertical flow pattern was followed, and water was not allowed to rise above the filter bed.

3.3. Monthly analysis of pilot scale CW effluent

Monthly analysis was required to understand the characteristics of the effluent. The results obtained for the monthly analysis of existing CW are tabulated in Table 3. The parameters such as turbidity and nitrates have exceeded the standard limits.

Table 3. Monthly analysis of wastewater.

Parameters Effluent Collected on	November	December	January	Drinking water standards[a]	Effluent standards[b]
pH	8.05	7.92	7.68	6.5-8.5	6.5-9
Temperature (°C)	29	29	29	--	--
Turbidity (NTU)	3	4	6	1	2-5
Dissolved oxygen (mg/l)	5.0	2.3	0	--	--
BOD₅ (mg/l)	0	16	20	0	30
Alkalinity (mg/l of CaCO₃)	140	185	179	200	--
Acidity (mg/l of CaCO₃)	105	125	115	--	--
Chlorides (mg/l)	44.99	52.5	37.5	250	--

Total dissolved solids (mg/l)	157.3	178	201	500	2000
Nitrates (mg/l)	0.89	1.02	2	0	---

[a]IS 10500: 2012 (drinking water standards).
[b]Standards and guidelines KSPCB 2016.

3.4. Performance analysis of pilot scale CW system using *Canna indica*

The purifying efficiency of the pilot scale wetland is calculated as shown in Table 4.

Table 4. Performance analysis.

Parameters	Influent (sewage)	Effluent (monthly average)	Efficiency (%)
pH	8.05	7.88	
Temperature ($^{\circ}$C)	29	29	0
Turbidity (NTU)	8	4.33	45.87
Dissolved oxygen	0	2.43	
BOD5 (mg/l)	185	12	93.51
Alkalinity (mg/l of $CaCO_3$)	1135	168	85.19
Acidity (mg/l of $CaCO_3$)	260	115	55.77
Chlorides (mg/l)	67.5	44.99	33.35
Nitrate (mg/l)	63.65	1.3	97.95
Total dissolved solids (mg/l)	608	178.77	70.59

3.5. Contribution of filter medium and plant medium

The results regarding the efficiencies of plant medium and filter medium reported by Yang et al. (2007) were taken as the secondary data for finding the contribution of plant medium separately. Tables 5 and 6 give the values of efficiencies in COD_{cr} and BOD_5 removal of control medium and *C. indica* (percentage error has been neglected).

Table 5. Percentage removal of COD_{cr} by the control and vegetated wetland (Yang et al., 2007).

Month (1)	Influent (mg/l) (2)	Control (%) (3)	C. indica and filter medium (%) (4)
May	185.03	67.90	88.53
June	152.28	74.39	81.45
August	46.67	30.14	44.33
September	186.63	81.76	85.95
October	201.62	84.28	99.20
November	215.78	94.39	96.19
December	506.74	85.33	80.49
Mean	213.54	74.03	82.31

Table 6. Percentage removal of BOD_5 by the control and vegetated wetland (Yang et al., 2007).

Month (1)	Influent (mg/l) (2)	Control (%) (3)	C. indica and filter medium (%) (4)
May	111.02	81.27	93.31
June	91.37	82.93	87.64
August	28.00	53.43	62.89
September	111.98	87.84	90.64
October	120.97	90.83	99.53
November	129.47	96.73	97.78
December	304.02	91.44	88.62
Mean	128.12	83.50	88.63

The contribution of *C. indica* in percentage is obtained by the following calculations: the percentage removal of COD_{cr} by filter media was 74.03. The percentage removal of COD_{cr} by plant media and the filter media was 82.31. Therefore, the percentage removal of COD_{cr} by plant media will be 8.28. similar calculations were done for BOD_5.

3.6. Phytoremediation modeling

A predictive model based on regression equation is developed using IBM SPSS taking day, temperature, humidity, phosphates, and pH as independent variables. The regression equation for predictive model for water quality parameters is shown in Table 7. The data of temperature and humidity was collected from www.timeanddate.com.

Table 7. Regression equations for predictive model.

Parameter	Regression equation	R^2
TDS	$-10.24*D - 0.84*T - 127.871*H + 360.073*pH - 79.813*P - 31.090*TKN - 1203.561$	0.638
BOD	$-0.376*D - 0.171*T - 6.268*H + 1.826*pH - 3.768*P - 0.865*TKN + 19.679$	0.785
COD	$0.208*D + 3.369*T + 104.807*H - 21.105*pH + 21.413*P + 13.241*TKN - 64.174$	0.894

In the above equations, the symbols stand for the following parameters: D: no. of days; T: temperature of sample; H: humidity; pH: potential of hydrogen; P: phosphate content; TKN: total Kjeldahl nitrogen. From comparison of actual and predicted values as shown in Figure 1, it can be observed that there has significant relation between actual and predicted values. The predicted values are nearly in the range of actual observed value.

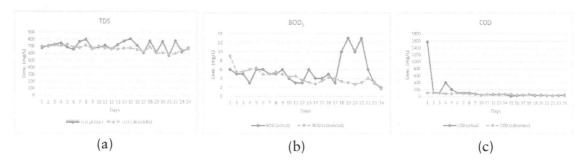

(a)　　　　　(b)　　　　　(c)

Figure 1. *Comparison between actual and calculated values of (a) TDS, (b) BOD_5, and (c) COD.*

4. CONCLUSION

The study analyzed the performance of the subsurface vertical flow pilot scale CW near Ladies Hostel, College of Engineering Trivandrum for the treatment of wastewater. The tests show that the pilot scale CW has 93% efficiency in BOD_5 removal, 70% efficiency in total dissolved solids removal, and 45% efficiency in removal of turbidity. Treated wastewater discharge meets the effluent standard and can be used for flushing, irrigation purpose. Also, the efficiency of CW system to treat wastewater using *Canna indica* has been studied. Results obtained from secondary data implies that the filter medium (percentage removal of BOD_5 by filter medium = 83%) takes up the non-toxic and/or biodegradable contaminants via various decomposition process taking place. Importantly, the plant medium (the percentage removal of COD_{cr} by plant medium = 8%) takes up and sequestrates the toxic and/or non-biodegradable contaminants as well as utilizes non-toxic and/or biodegradable contaminants. As the values of coefficient of determination (R^2) show the dependent and independent variables are approximately linearly dependent. Also, the predicted values using regression model are comparable to the range of actual values. Therefore, CW can be made a good option for water treatment and reuse.

REFERENCES

[1] Cui, L., Ouyang, Y., Lou, Q., Yang, F., Chen, Y., Zhu, W., and Luo, S., "Removal of nutrients from wastewater with *Canna indica* L. under different vertical-flow constructed wetland conditions," *Ecol Eng*, vol. 36, pp. 1083–1088, 2010.

[2] Haritash, A. K., Sharma, A., and Bahel, K., "The potential of Canna lily for wastewater treatment under Indian conditions," *Int J Phytorem*, vol. 17, pp. 999–1004, 2015.

[3] Herath, I., and Vithanage, M., "Phytoremediation in constructed wetlands," *Phytoremediation: Manag Environ Cont*, vol. 2, pp. 243–263, 2015.

[4] Jacome, J. A., "Performance of constructed wetland applied for domestic waste water treatment: Case study at Boimorto (Galicia, Spain)," *J Ecol Eng*, vol. 95, pp. 324–329, 2016.

[5] Joseph, T., "Design and development of constructed wetland system for CET Ladies Hostel," M. Tech. Thesis, APJ Abdul Kalam Technological University, Trivandrum, India, 2018.

[6] Ramaprasad, C., and Philip, L., "Surfactants and personal care products removal in laboratory scale horizontal and vertical flow constructed wetlands while treating greywater," *J Chem Eng*, vol. 284, pp. 458–468, 2016.

[7] Sandoval, L., Zamora-Castro, S. A., Vidal-Alvarez, M., and Marin-Muniz, J. L., "Role of wetland plants and use of ornamental flowering plants in constructed wetlands for wastewater treatment: A review," *Appl Sci*, vol. 9, article no: 685, 2019.

[8] Sugyana, K., and Sebastian, P. S., "Phytoremediation prospective of Indian shot (*Canna indica*) in treating the sewage effluent through hybrid reed bed (HRB) technology," *Int J Chem Stud*, vol. 5, pp. 102–105, 2017.

[9] Tilley, E., Ulrich, L., Luthi, C., Reymond, Ph., and Zurbrugg, C., *Compendium of Sanitation Systems and Technologies*, 2nd Edition, Eawag: Swiss Federal Institute of Aquatic Science and Technology, 2014.

[10] Vidali, M., "Bioremediation. An overview," *Pure Appl Chem*, vol. 73, pp. 1163–1172, 2001.

[11] Vymazal, J., "Constructed Wetlands for Wastewater Treatment," *Water*, vol. 2, pp. 530–549, 2010.

[12] Wang, M., Zhang, D. Q., Dong, J. W., and Tan, S. K., "Constructed wetlands for wastewater treatment in cold climate – A review," *J Environ Sci*, vol. 10, pp. 1–19, 2017.

[13] Yang, Q., Chen, Z., Zhao, J., and Gu, B., "Contaminant removal of domestic wastewater by constructed wetlands: Effects of plant species," *J Integr Plant Biol*, vol. 49, pp. 437–446, 2007.

[14] Zhang, Z., Rengel, Z., and Meney, K., "Removal of nutrients from secondary-treated municipal wastewater in wetland microcosms using ornamental plant species," *Int J Environ Waste Manag*, vol. 1, pp. 363–375, 2007.

12. Biomethanation Potential of Organic Fraction of Municipal Solid Waste (OFMSW) Using Co-Digestion

Anjali M and Swarnalatha K

Department of Civil Engineering, College of Engineering, Trivandrum, India

*Corresponding author: anjali.eer@gmail.com

ABSTRACT: The use of "waste to energy" technology is one of the best ways to accomplish sustainable energy development, and this is becoming well-established on a global scale. The most common method is anaerobic digestion, which transforms organic materials into clean, renewable products. Increasing demand of energy and non-feasible anaerobic digestion process drives to necessity of rapid development in anaerobic co-digestion technology especially for food waste. Moreover, parameter optimization for different substrate combinations is very important to limit the challenges of food waste for improved biomethanation potential. The present study intends to evaluate biomethanation potential of organic fraction of municipal solid waste (OFMSW) using co-digestion. For enhancing biogas production, OFMSW is co-digested with sewage sludge and cow dung. Maximum cumulative biogas yield of 37.47 ml/gVS observed under optimum ratio of FW:CD (60:40) for a 14 days retention time.

KEYWORDS: Anaerobic Co-digestion, OFMSW, Biomethanation Potential

1. INTRODUCTION

Municipal solid waste (MSW) management and energy insecurity are the bottleneck problem facing on a global scale. About 1.5 lakh tonnes of MSW are generated per day in Urban India, of which only 70–75% gets collected and 20–25% is treated (Gosh et al., 2020). Also, the collected un-segregated waste is majorly dumped in un-engineered landfills leading to anthropogenic methane emissions which promote climate change and severe environmental and groundwater contamination (Negi, 2018). Food and green waste make up the majority of waste on a global scale, accounting for 44% of all waste (Gosh et al., 2020). In the Indian scenario, biodegradable waste comprises 51% of total municipal solid waste generated (Kasinath et al., 2021). Four million tons of municipal solid waste is generated annually in Kerala (Suchitwa Mission, 2020). Consumption and waste management decisions made by individuals and governments have an impact on the daily health, output, and cleanliness of communities. Poor waste management is hindering economic growth, damaging the world's oceans, clogging sewers and causing flooding, spreading disease, increasing respiratory issues, injuring animals who inadvertently consume waste, and more. Action must be taken immediately at all societal levels to address the unmanaged and badly managed trash from decades of economic growth.

Renewable alternative solutions to this problem are conversion of waste to energy, which is being practised by many countries. Anaerobic digestion is the method for biological decomposition of organic matter without oxygen and producing biogas from waste, which mainly contains methane and carbon dioxide. Utilizing the latent energy in organic matter by biomethanation or anaerobic decomposition of biodegradable waste. It creates slurry, a good organic fertilizer, and biogas, which can replace gaseous fossil fuels like LPG, CNG, etc. and lower greenhouse gas emissions (Suchitwa Mission, 2020).

Chapter 12 DOI- 10.1201/9781032657271-12

Anaerobic co-digestion offers a way to simultaneously digest two or more feedstocks, so overcoming the limitations of mono digestion (Bedoi et al., 2020). The main advantages of co-digestion are improved system stability and methane yield due to the synergistic effects of encouraging a more diverse microbial community, better nutrient balance (proper carbon-to-nitrogen [C/N] ratio and supplementation of trace elements), improved buffering capacity, dilution of toxic compounds, including heavy metals, safe and better quality digestate for agricultural applications, and reduction of antibiotic resistance genes (Karki, 2021). The study gives a holistic approach for efficient anaerobic co-digestion (Aco-D) of FW and substrates such as sewage sludge, and cow dung to provide a comprehensive result.

2. MATERIALS AND METHODOLOGY

2.1. Sample collection

Food waste was collected from Hostel canteen, College of Engineering Trivandrum. The food waste (FW) is composed of different leftover of cooked foods, vegetable peelings, fruit waste, etc. *Collected samples were homogenized by grinding into small size for the better performance of anaerobic digestion. Sewage sludge (SS) was collected from Muttathara wastewater treatment plant, Trivandrum.* Cow dung (CD) was collected from a cow shed near College of Engineering Trivandrum. It was obtained in solid form and then homogenized. Grinded samples were stored separately in the freezer until needed. The characterization study such as pH, moisture content, total solids, total volatile solids, C/N ratio, etc. carried out using standard procedure.

2.2. Experimental setup

Co-digestion test of different substrates was carried out using water displacement method. The bioreactor setup consists of 3 glass bottles (Figure 1). Glass digesters of volume 500 ml, 1000 ml, and 500 ml, respectively, were fabricated in the laboratory for study. Out of which amber color bottle is an anaerobic digester, 1000 ml bottle is used for storage of water for a working volume of 800 ml and 200 ml head space.

Figure 1. Test set up for anaerobic co-digestion.

Once the gas formation started in the anaerobic digester, the gas passed into the measuring bottle displaces the water into the third bottle and occupy the space. The volume displaced was noted from the scale of the measuring bottle.

2.3. Optimization of parameters

Experimental studies include optimization of parameters for anaerobic digestion such as substrate-to-co-substrate ratio and optimization of pH. Co-digestion test was carried out on a volume-based analysis. Tests were conducted at room temperature, initial pH was maintained at 7. Food waste and sewage sludge were fed into 500 ml batch reactor in different ratios and biogas volume generated on a daily basis was measured for a retention time of 10 days. Six batch anaerobic digesters namely AD1, AD2, etc. were used for co-digestion test. To analyze the optimum concentration of food waste to sewage sludge, six batch experiments (100% food waste, 100% sewage sludge, and with different food waste to sewage sludge ratio as 20:80, 40:60, 60:40, 50:50, 80:20) were conducted in 500 ml anaerobic digesters. Optimization of FW and CD is also done following the same procedure.

3. RESULTS AND DISCUSSION

3.1. Characterization study

The food waste collected was acidic in nature, pH obtained was about 5.4. Sewage sludge and cow dung have a neutral pH value, in the range of 6.8–7.6. Moisture content of each sample was more than 80%, which is desirable for anaerobic digestion process. The main characteristics required for anaerobic digestion process are C/N content. Maximum content of organic carbon was found in cow dung (40.1%), food waste (39.02%), and sewage sludge (16.4%) (Table 1). The content of total nitrogen is varying from 2.01% (cow dung), 2.15% (food waste), 2.17% (sewage sludge), 3.1% (sewage sludge) to 3.2% (water hyacinth). The results obtained agreed with the literature data (Karki, 2021).

Table 1. Charecterization study.

Sl no.	Sample name	Moisture content (%)	Total solids (%)	Volatile solids (%)	pH	C (%)	N (%)	C/N
1	Food waste	94.3	25	22	5.4	39.02	2.15	18.15
2	Sewage sludge	68.9	6.1	4.4	7.4	16.4	2.17	7.55
3	Cow dung	69.4	24.8	19	7.5	40.1	2.01	19.95

3.2. Co-digestion of food waste and sewage sludge

Biogas production was observed by conducting optimization of food waste to sewage sludge ratio for a retention time of 10 days and total volume of 935 ml obtained. Maximum biogas yield of 25.47 ml/gVS obtained for food waste: sewage sludge ratio 60:40. Optimized ratio for co-digestion for maximum biomethanation potential is 60:40. Result observed was in accordance with the literature (Gosh et al., 2020). Performance of anaerobic co-digestion is dependent on C/N ratio of feedstocks (Gosh et al., 2020). Inappropriate C/N ratio of single substrate is not optimum for anaerobic co-digestion. Higher biogas yield was obtained for a C/N ratio of 15.28 which is found optimal (Table 2). Cumulative biogas yield of food waste alone is 10.44 mL/gVS, while food waste co-digested with sewage sludge resulted in cumulative biogas yield of 25.47 mL/gVS. Result showed that 143% increase of cumulative biogas yield obtained when food waste co-digested with sewage sludge. Figures 2 and 3 show daily biogas volume and cumulative biogas yield for co-digestion of FW and SS.

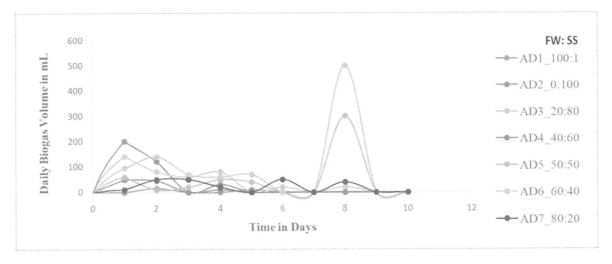

Figure 2. *Daily biogas yield for co-digestion of FWand SS.*

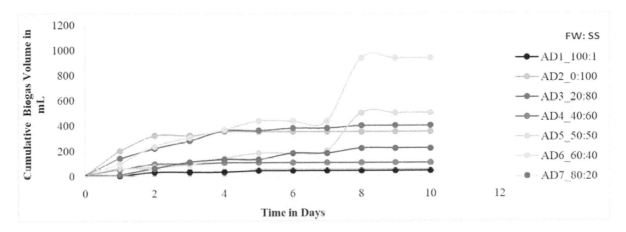

Figure 3. *Cumulative biogas yield for co-digestion of food waste: Sewage sludge.*

Table 2. Optimization of food waste and sewage sludge.

Substrates	Digester name	Food waste: Sewage sludge	Total solids (g/l)	Volatile solids (g/l)	Total biogas (ml)	Cumulative biogas yield (ml/gVS)	C/N (%)
Food waste	AD1	_	82	76.5	400	10.44	18.15
Sewage sludge	AD2	_	58	56.4	350	12.41	7.55
Co-digestion 1	AD3	20:80	62	58	400	13.79	9.66
Co-digestion 2	AD4	40:60	68	64.2	105	3.27	11.77
Co-digestion 3	AD5	50:50	74.7	69.3	480	13.85	12.82
Co-digestion 4	AD6	60:40	76.2	73.4	935	25.47	15.28
Co-digestion 5	AD7	80:20	79.8	72.8	220	6.04	15.11

Table 2 shows the optimization results of food waste and sewage sludge. It can be observed that mono digestion of food waste and sewage sludge gives less biomethanation yield than that of co-digestion.

3.3. Optimization of food waste and cow dung

Optimization of food waste to cow dung ratio was conducted for a retention time of 14 days and a total volume of 1900 ml obtained. Maximum biogas yield of 37.14 ml/gVS obtained for FW:CD ratio 60:40 (Figure 4). Optimized ratio for co-digestion for maximum biomethanation potential is 60:40. Maximum biogas yield obtained for a C/N ratio of 18.87. Cumulative biogas yield of food waste alone is 10.44 mL/gVS. While co-digestion of food waste along with cow dung resulted in 255% increase in cumulative biogas yield.

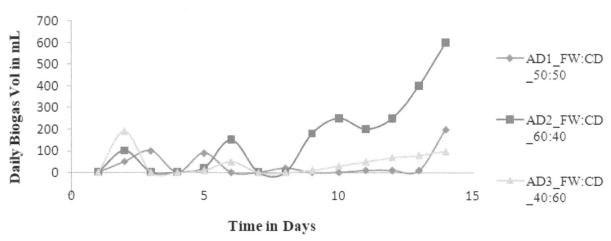

Figure 4. *Daily biogas volume for co-digestion of FW and CD.*

Table 3. Optimization of Food waste and Cow Dung.

Substrates	Digester name	FW:CD	Total solids (g/l)	Volatile solids (g/l)	Total biogas (ml)	Cumulative biogas yield (mL/gVS)	C/N (%)
Cow dung	AD1	–	69.2	69.4	400	11.52	19.95
Co-digestion 1	AD2	60:40	125	102.3	1900	37.14	18.87
Co-digestion 2	AD3	50:50	98	95.4	590	12.36	19.01
Co-digestion 3	AD4	40:60	96	94.6	400	8.45	19.25

Table 3 shows the parameters for anaerobic co-digestion, total biogas volume, and cumulative biogas yield of each anaerobic digester. Maximum biogas yield obtained from co-digestion 1, AD2.

4. CONCLUSIONS

The study revealed that co-digestion of food waste along with sewage sludge and cow dung has a significant role in biomethanation potential. Co-digestion experiment conducted for a retention time of 10–14 days. Co-digestion test was conducted among different combination of substrates. The best substrate combinations for obtaining maximum biomethanation potential have been determined. The highest biomethantion potential and biogas yield were ensured by the co-digestion of the following mixtures: Cumulative biogas yield of food waste alone is 10.44 mL/gVS, while food waste co-digested with sewage sludge resulted, cumulative biogas yield of 25.47 mL/gVS in the ratio 60:40. This means

143% increase obtained when food waste co-digested with sewage sludge. Food waste and cow dung biogas yield was 37.14 mL/gVS in the ratio 60:40, resulted a percentage increase of 255% in cumulative biogas yield to that of mono digestion of food waste. So it can be well analyzed that co-digestion really having a significant role in biomethanation potential of organic fraction of municipal solid waste best combination for co-digestion was food waste and cow dung.

REFERENCES

[1]   Amodeo, C., Hattou, S., Buffiere, P., and Benbelkacem, H., "Temperature phased anaerobic digestion (TPAD) of organic fraction of municipal solid waste (OFMSW) and digested sludge (DS): Effect of different hydrolysis conditions," *Waste Manag*, 2021.

[2]   Bedoi, Robert, Spehar, A., and Puljko, J., "Opportunities and challenges: Experimental and kinetic analysis of anaerobic co-digestion of food waste and rendering industry streams for biogas production," *Renew Sustain Energy Rev*, 2020.

[3]   Gao, M., Yang, M., Xie, D., Wu, C., and Wang, Q., "Effect of co-digestion of tylosin fermentation dreg and food waste on anaerobic digestion performance," *Biores Technol*, 2021.

[4]   Gosh, P., Kumar, M., Kapoor, R., Kumar, S., Sing, L., Vijay, V., and Takur, I., "Enhanced biogas production from municipal solid waste via co-digestion with sewage sludge and metabolic pathway analysis," *Biores Technol*, 2020.

[5]   Gosh, P., Kumar, M., Kapoor, R., Kumar, S., Sing, L., Vijay, V., and Takur, I., "Enhanced biogas production from municipal solid waste via co-digestion with sewage sludge and metabolic pathway analysis," *Biores Technol*, 2021.

[6]   Karki, Renisha, "Anaerobic co-digestion: Current status and perspectives," *Biores Technol*, 2021.

[7]   Kasinath, A., Ksiazek, S., Szopinska, M., Bylinski, H., Artichowicz, W., Skwarek, A., "Biomass in biogas production: Pretreatment and codigestion," *Renew Sustain Energy Rev*, 2021.

[8]   Mehariya, S., Patel, A. K., Obulisamy, P. K., Punniyakotti, E., and Wong, J. W. C., "Co-digestion of food waste and sewage sludge for methane production: Current status and perspective," *Biores Technol*, 2018.

[9]   Negi, H. D., "Biomethanation potential for co-digestion of municipal solid waste and rice straw: A batch study," *Biores Technology*, 2018.

[10] Suchitwa Mission, *Introduction And Strategic Enviromental Assessment Of Waste Management Sector In Kerala*, 2020.

13. Comparative Lifecycle Assessment of Alcohol-Based Hand Sanitizers Using Open LCA

Vishnudatha V, Priyanka, Benita Nishil, Catherine Chakkalakkal and Ambika S*
Environmental Engineering Division, Department of Civil Engineering, Indian Institute of Technology Hyderabad (IITH), Kandi, Sangareddy, Telangana 502285, India.
*Corresponding author: ambika@ce.iith.ac.in

ABSTRACT: During the COVID-19 pandemic, the usage of hand sanitizers increased significantly as a crucial preventive measure against the spread of the virus. In response to the soaring demand, the production capacity of hand sanitizers in India was expanded by 1000 times. However, concerns arose regarding the potential environmental impacts associated with this rapid increase in production. Thus, this study aimed to assess the environmental consequences of ethanol and isopropanol-based hand sanitizers through a comparative life cycle assessment (LCA) using the Open LCA software. The assessment considered various impact categories, including acidification potential, eutrophication, climate change, aquatic toxicity, human toxicity, ozone layer depletion, and terrestrial ecotoxicity. The impact assessment methods utilized were CML baseline and ReCipe endpoint H. The findings from the midpoint-based impact assessment using the CML baseline method revealed that ethanol-based hand sanitizer exhibited higher acidification potential, eutrophication, aquatic toxicity, human toxicity, ozone layer depletion, and terrestrial ecotoxicity compared to isopropanol-based hand sanitizer. Conversely, isopropanol-based hand sanitizer demonstrated higher global warming potential, depletion of abiotic resources (fossil fuels), and photochemical oxidation potential. Furthermore, the ReCipe endpoint H results indicated that ethanol-based hand sanitizer had more pronounced adverse effects on the ecosystem and human health, except for photochemical oxidant formation, where the effects of isopropanol-based hand sanitizer were predominant. Additionally, the analysis showed that the effects related to climate change were similar for both ethanol and isopropanol-based sanitizers. In summary, the study highlights that ethanol-based hand sanitizer poses a higher overall environmental impact in terms of various impact categories. Conversely, isopropanol-based hand sanitizers exhibit a comparatively lower impact, particularly in terms of global warming potential, depletion of abiotic resources, and photochemical oxidation potential.

KEYWORDS: Hand Sanitizer, Life Cycle Assessment, Open LCA Software

1. INTRODUCTION

1.1. General

Hand sanitizer, also known as hand disinfectant, is available in liquid, foam, and gel forms and is primarily used to eliminate microorganisms such as bacteria and virus (*Boyce and Pittet, 2002;* Hand sanitizer, 2017). There are two types of hand sanitizers: alcohol-based and alcohol-free. Alcohol-based sanitizers typically contain 70–85% ethanol or isopropanol, while alcohol-free variants incorporate antimicrobial agents like triclosan. The effectiveness of sanitizers depends on factors such as exposure time, quantity used, and frequency of application (Barton, 2012). The usage of hand sanitizers has

Chapter 13 DOI- 10.1201/9781032657271-13

witnessed a sharp increase due to the COVID-19 pandemic, resulting in a significant rise in production since 2019. The World Health Organization (WHO) guidelines endorsing the use of alcohol-based hand sanitizers among healthcare workers and in areas with water scarcity, recognizing their effectiveness in combating the virus had greatly contributed to the market growth of hand sanitizers (2010). The COVID-19 pandemic witnessed a significant surge in hand sanitizer production capacity in India, increasing it by 1000 times compared to pre-pandemic levels (from 10 lakh litres per annum to 30 lakh litres per day) (Jayan et al., 2021). As a result, India has become the second-largest consumer of hand sanitizer globally, following the United States (Hand sanitizer market). Given this substantial increase in production, it is crucial to evaluate the environmental implications associated with the heightened use of ethanol and isopropanol-based hand sanitizers. This study focuses on assessing the impact of these hand sanitizers on the environment. The objectives of conducting a life cycle assessment (LCA) on alcohol-based hand sanitizers include comparing the environmental impacts of ethanol-based and isopropanol-based hand sanitizers, assessing, and quantifying the effects of hand sanitizer production using Open LCA software, evaluating the differences in impacts between ethanol and isopropanol-based hand sanitizers, and examining the potential environmental consequences of the rapid increase in hand sanitizer production. The study aims to evaluate the environmental performance of alcohol-based hand sanitizers using a cradle-to-gate approach, which encompasses the entire life cycle from resource extraction to compounding. It seeks to provide a quantitative assessment of the environmental impact of alcohol-based hand sanitizers, serving as a reference point for their life cycle performance. Additionally, the study intends to conduct a comparative LCA between ethanol and isopropanol-based hand sanitizers.

2. METHODOLOGY

2.1. LCA methodology

This study presents a quantitative reference for the LCA of alcohol-based hand sanitizer, considering various phases. The LCA focused on the primary raw materials used in the production, such as ethanol, hydrogen peroxide, and glycerol. Additionally, the production of Polyethylene terephthalate bottles was also considered. The assessment adopted a cradle-to-gate approach, which encompasses the entire product life cycle from raw material extraction to the factory gate. The functional unit, which provides a quantified description serving as a reference for impact assessment calculations and product comparisons, was set as 1000 bottles of sanitizer, each with a capacity of 1 l (Arzoumanidis et al., 2019). The analysis considered the same functional unit for both ethanol-based and isopropanol-based hand sanitizers. The LCA study included inputs such as ethanol or isopropanol, glycerol, hydrogen peroxide, and water for sanitizer production. Polyethylene, polypropylene, and paper were considered for the production of Polyethylene terephthalate bottles. The inputs used for the analysis are given in Tables 1 and 2.

2.2. Impact assessment method

CML baseline and ReCipe Endpoint H were used as the impact assessment methods (OpenLCA). Midpoint impacts obtained using CML baseline such as climate change, acidification potential, terrestrial ecotoxicity, photochemical oxidation, marine aquatic ecotoxicity, eutrophication, human toxicity, ozone layer depletion, freshwater aquatic ecotoxicity, and depletion of abiotic resources were considered for the LCA analysis (Guinée, 2002). Endpoint impacts were analyzed using ReCipe Endpoint H method. It takes into account damage caused to health, damage caused to ecosystem, and damage to resource availability (Guinée, 2002).

2.3. Study assumptions and limitations

The study focused on the assessment of liquid-type alcohol-based hand sanitizer. In the analysis, it was assumed that a single Polyethylene terephthalate bottle used for packaging the sanitizer weighed 20 g (Marathe et al., 2017). To produce the sanitizer, only the primary ingredients specified in the WHO formulations were considered, and fragrance materials and colorants were not included (WHO, 2010). The distance between the point of production of Polyethylene terephthalate bottles and the sanitizer production unit was assumed to be 200 km. Additionally, it was assumed that the distributed sanitizer would be distributed within a radius of 100 km.

Table 1. Inputs used for producing a single Polyethylene terephthalate bottle of 1 l.

	Compounds	Percentage v/v	Weight (g)	Reference
Caps	Polypropylene	2.5	0.5	Foolmaun and Ramjeeawon (2012)
	High-density polyethylene	2.5	0.5	Foolmaun and Ramjeeawon (2012)
Labels	Polypropylene	0.5	0.1	Foolmaun and Ramjeeawon (2012)
	Paper	0.5	0.1	Foolmaun and Ramjeeawon (2012)
Bottle	Polyethylene terephthalate	94	18.8	Foolmaun and Ramjeeawon (2012)

Table 2. WHO recommended hand sanitizer formulations for production of 1 l of sanitizer.

Ethanol-based sanitizer		Isopropanol-based sanitizer		References
Ethanol 96%	833.3 ml	Isopropanol 99%	751.5 ml	WHO (2010); Roy and Dutta (2019)
Hydrogen peroxide 3%	41.7 ml	Hydrogen peroxide 3%	41.7 ml	WHO (2010)
Glycerol 98%	14. 5 ml	Glycerol 98%	14.5 ml	WHO (2010); Ghannadzadeh and Tarighaleslami (2020)
Water	110.5 ml	Water	192.3 ml	WHO (2010)

3. RESULTS AND DISCUSSION

3.1. Results of Open LCA

The impact analysis conducted using the CML baseline impact assessment method unveiled the effects of ethanol-based hand sanitizer across 11 distinct categories, including acidification potential, terrestrial ecotoxicity, climate change (GWP 100), photochemical oxidation, marine aquatic ecotoxicity, eutrophication, human toxicity, depletion of abiotic resources, freshwater aquatic ecotoxicity, and ozone layer depletion. Notably, sanitizer production contributed significantly (97.66%) to the acidification potential, whereas bottle production accounted for a minor fraction (2.34%). Among the inputs, ethanol usage in sanitizer production emerged as the primary contributor to acidification, primarily due to the release of SO_2 during the fermentation of sugarcane and molasses to produce ethanol. Conversely, the production of glycerol had a greater impact on global warming when compared to ethanol production. Additionally, ethanol production was responsible for 86% of the eutrophication potential.

Similarly, the impact analysis conducted on isopropanol-based hand sanitizer, utilizing the CML baseline impact assessment method, encompassed the same 11 impact categories. In this analysis, the production of isopropanol exhibited a prominent contribution to impact categories such as acidification potential and climate change. Conversely, the production of glycerol played a substantial role in impact categories such as terrestrial ecotoxicity, eutrophication, ozone layer depletion, human toxicity, photochemical oxidation, and freshwater aquatic ecotoxicity. The comprehensive results of the impact analysis for both ethanol and isopropanol-based hand sanitizers, employing the CML baseline method in Open LCA software, can be found in Table 3.

Table 3. Results of impact analysis of ethanol and isopropanol-based hand sanitizer using CML baseline method in Open LCA software.

Name of the category	Ethanol-based sanitizer	Isopropanol-based sanitizer	Unit
Terrestrial ecotoxicity—TETP inf	233.48873	175.88515	Kg 1,4-dichlorobenzene eq.
Acidification potential—average Europe	11.65653	4.78728	Kg SO2 eq.
Climate change—GWP100	939.50715	1385.23252	Kg CO2 eq.
Photochemical oxidation—high NOx	0.49431	0.60469	Kg ethylene eq.
Marine aquatic ecotoxicity—MAETP inf	4.74515E5	2.81168E5	Kg 1,4-dichlorobenzene eq.
Eutrophication—generic	7.57098	1.38678	Kg PO4 eq.
Human toxicity—HTP inf	717.64185	216.78948	Kg 1,4-dichlorobenzene eq.
Depletion of abiotic resources—elements, ultimate reserves	0.00167	0.00044	Kg antimony eq.
Freshwater aquatic ecotoxicity—FAETP inf	811.07032	465.91154	Kg 1,4-dichlorobenzene eq.
Depletion of abiotic resources—fossil fuels	7933.53656	3.12925E4	MJ
Ozone layer depletion—ODP steady state	6.24551E-5	3.45577E-5	Kg CFC 11 eq.

3.2. Comparative analysis of the result

The comparative analyses conducted in the Open LCA Software provided valuable insights into the environmental and health impacts of ethanol and isopropanol-based hand sanitizers. The results from the first analysis using CML baseline method (Figure 1a) showed that ethanol-based hand sanitizer had significant contributions to impact categories such as acidification potential, depletion of biotic resources, eutrophication, aquatic toxicity, human toxicity, ozone layer depletion, and terrestrial ecotoxicity. In contrast, isopropanol-based hand sanitizer had a higher impact on categories including climate change, depletion of abiotic resources (fossil fuels), and photochemical oxidation. The second analysis using the ReCipe Endpoint H impact assessment method (Figure 1b) focused on the effects of the two sanitizers on ecosystem and human health. It revealed that ethanol-based hand sanitizer had a greater overall impact on ecosystem and human health, except for the category of photochemical oxidant formation, where isopropanol-based hand sanitizer had a more dominant effect. Interestingly, the analysis indicated that the effects of climate change were similar for both ethanol and isopropanol-based sanitizers. Overall, these comparative analyses provide a comprehensive understanding of the environmental and health impacts associated with the use of ethanol and isopropanol-based hand sanitizers, highlighting the specific areas where each type has a significant influence.

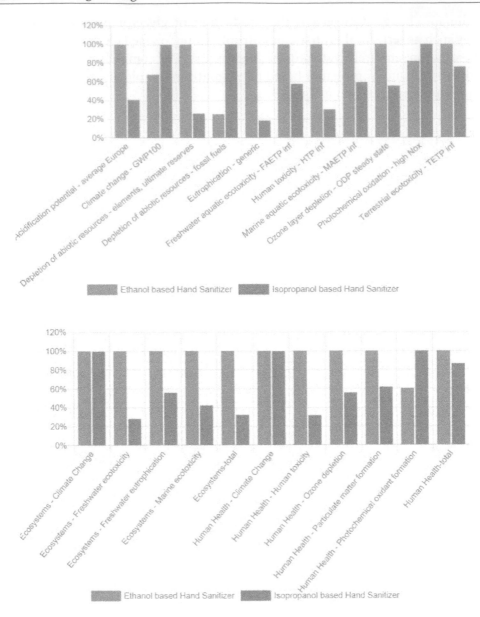

Figure 1. *(a) Comparative analysis of results of CML baseline method, (b) comparative analysis of results of ReCipe Endpoint H method.*

4. CONCLUSION

The comparative LCA conducted in this study revealed the environmental impact differences between ethanol and isopropanol-based hand sanitizers. Using the CML baseline method for midpoint-based impact assessment, ethanol-based hand sanitizer exhibited higher acidification potential, eutrophication, aquatic toxicity, human toxicity, ozone layer depletion, and terrestrial ecotoxicity. On the other hand, isopropanol-based hand sanitizer showed higher global warming potential, depletion of abiotic resources (fossil fuels), and photochemical oxidation potential. Furthermore, the contribution tree analysis showed that the production of ethanol had a greater impact on most categories, except for global warming where glycerol was the major contributor. Utilizing the ReCipe endpoint method, it was found that ethanol-based hand sanitizer had more significant effects on the ecosystem and human health, except for photochemical oxidant formation, where the effects of isopropanol-based hand sanitizer dominated. Notably, the effects of climate change were found to be comparable between ethanol and isopropanol-based sanitizers. In conclusion, the study demonstrates that ethanol-based hand sanitizer has a higher overall environmental impact, while isopropanol-based hand sanitizers exhibit a comparatively lower overall impact.

REFERENCES

[1] Aitor, A., Rodriguez, P., and Ciroth, C., "Impact assessment methods in life cycle assessment and their impact categories," *Green Delta*, no. January 2014, pp. 1–23, 2014.

[2] Arzoumanidis, I., D'Eusanio, M., Raggi, A., and Petti, L., "Functional unit definition criteria in life cycle assessment and social life cycle assessment: A discussion," *Perspectives on Social LCA*, pp. 1–10, 2019.

[3] Barton, L., "Hand sanitisers: Saved by the gel?" *The Guardian*, 13 May 2012. https://www.theguardian.com/society/2012/may/13/do-we-really-need-hand-sanitisers

[4] *Boyce JM, and Pittet D*, "Guideline for hand hygiene in health-care settings. Recommendations of the healthcare infection control practices advisory committee and the HICPAC/SHEA/APIC/IDSA Hand Hygiene Task Force. Society for Healthcare Epidemiology of America/Association for Professionals in Infection Control/Infectious Diseases Society of America," *MMWR Rec Rep, vol. 51, no. RR-16, pp. 1–45, quiz CE1–4, 2002*. PMID 12418624.

[5] Ciroth, A., Noi, C., Lohse, T., and Srocka, M., "OpenLCA 1.10 – Comprehensive user manual," no. V 1.10.2, p. 127, 2020. https://www.openlca.org/learning/

[6] Foolmaun R. K., and Ramjeeawon, T. "Disposal of post-consumer polyethylene terephthalate (PET) bottles: Comparison of five disposal alternatives in the small island state of Mauritius using a life cycle assessment tool," *Environ Technol*, vol. 33, no. 5, pp. 563–572, 2012.

[7] Ghannadzadeh, A., and Tarighaleslami, A. H., "Environmental life cycle assessment of glycerine production: Energy transition from natural gas to biomass," *Sustain Energ Technol Asses*, vol. 42, p. 100775, 2020.

[8] Guinée, J. B., *Handbook on Life Cycle Assessment: Operational Guide to the ISO standards*. Dordrecht: Kluwer Academic Publishers, 2002.

[9] Hand sanitizer – definition of hand sanitizer in English. Oxford Dictionaries, *2017*. Retrieved 12 July 2017.

[10] Hand sanitizer market – growth, trends, COVID-19 impact, and forecasts (2021–2026).

[11] Harding, K. G., "A technique for reporting life cycle impact assessment (LCIA) results," *Ecol Ind*, vol. 34, pp. 1–6, 2013.

[12] Jayan, T. V., "India's hand sanitizer production capacity grew 1,000 times during covid-19 pandemic," *The Hindu Business Line*, 09 Jul 2021. https://www.thehindubusinessline.com/markets/commodities/indias-hand-sanitizer-production-capacity-grew-1000-times-during-covid-19-pandemic/article34832010.ece.

[13] Jing, J. L., Pei Yi, T., Bose, R. J., McCarthy, J. R., Tharmalingam, N., and Madheswaran, T., "Hand sanitizers: A review on formulation aspects, adverse effects, and regulations," *Int J Environ Res Pub Heal*, vol. 17, no. 9, p. 3326, 2020.

[14] Marathe, K. V., Chavan, K., and Nakhate, P., "Life cycle assessment (LCA) of polyethylene terephthalate (PET) bottles – Indian perspective," p. 53, 2017.

[15] OpenLCA Nexus. https://nexus.openlca.org/database

[16] Roy, P., and Dutta, A., "Life cycle assessment (LCA) of bioethanol produced from different food crops: Economic and environmental impacts," *Bioethan Prod from Food Crops*, pp. 385–399, 2019.

[17] WHO, "Guide to local production: WHO-recommended hand rub formulations introduction," WHO, no. April, p. 9, 2010. https://www.who.int/gpsc/5may/Guide_to_Local_Production

14. Emergy-Based Approaches to Assess the Impact Caused to Various Ecosystem Services Resulting from Natural Resource Extraction

Bincy S* and Praveen A

Department of Civil Engineering, Rajiv Gandhi Institute of Technology, APJ Abdul Kalam Technological University, Kerala, India

*Corresponding author: bincyneema@gmail.com

ABSTRACT: Sustainability in resource supply is very essential to ensure stable operations of construction sector in a developing economy. The excessive use of non-renewable energy, the over exploitation of natural materials, and the exhaustion of abiotic mineral resources are the major problems often linked with the non-availability of resources to support the construction industry. As the majority of resources used in the above-mentioned sector are directly pooled locally, geographically linked parameters for environmental evaluations are needed to forecast the availability of natural materials and to assess the technological choice for resource supply. Among the various techniques developed which utilize thermodynamic principles for evaluating environmental sustainability, the emergy analysis includes environmental work provided by the biosphere together with the energy used up in the formation of resources. Emergy referred as "energy memory" quantifies previously used-up available energy in the formation of a product on a common basis of solar equivalent energy expressed as solar emjoule (sej). Tracing all resource inputs and depreciated solar equivalent energies for the contribution of a resource all entropy losses in its production are accounted for in emergy evaluations. The study proposes a reliable indicator that could assess the consumptive pattern of resources in the construction sector at a regional level. The variation of available emergy with cost of the most widely used construction materials for 25 years is mapped to elicit the time scale impact of resource use patterns using the data collected from published sources. The mapped models indicate the ecosystem resilience and its ability to stabilize and provide a regular supply of resources. Emergy increase in construction leads to emergy loss in delivering the ecosystem services. The environmental services lost with sand extraction namely groundwater recharge, nutrient building, and loss in ecosystem productivity are accounted using emergy approach which helps formulate regulatory strategies for long-term resource extraction

KEYWORDS: Emergy, Sustainability, Ecosystem, Natural Resources

1. INTRODUCTION

Rising demand for natural resources has been a cause for the degradation of the ecosystem which is mostly triggered at the regional level and later manifested to the national and global level in terms of loss of biodiversity, ecosystem degradation, and rising climate change. Depletion of bulk resources is observed to be negligible at the earth or world scale, at a country scale often depletion reported as low and at a regional scale depletion is obvious (Habert et al., 2010). Therefore, appropriate geographic representations for environmental parameters used for evaluations are needed to forecast the availability of natural materials and to assess the technological choice for resource supply.

Chapter 14 DOI- 10.1201/9781032657271-14

Utility of thermodynamic principles for effective evaluation of the sustainability of building material use is widely established. Among the various tools developed to evaluate energy, material requirements and environmental impact of these resources like life cycle analysis, emergy analysis, exergy analysis, ecological footprint analysis, etc. only emergy analysis provides an extensive judgment of natural investment and human potential. Emergy is defined as the total amount of solar energy that is used directly or indirectly to generate a service or product expressed in solar equivalent joules (sej) (Odum, 1996) which is calculated by multiplying its available energy inflow by respective transformities. The unit emergy values (UEV) or emergy intensities or transformities is the emergy required to produce one unit of a product or service, expressed per unit basis of energy, money, time, mass, etc. (Brown and Ulgiati, 2004). The emergy baseline, used as a reference for transformity calculations is derived from available energy supporting the biosphere viz, solar radiation, geothermal sources, and dissipation of tidal momentum.

Methods to quantify the timescale effect of natural resource consumption are not available. As the depletion of natural resources increases, the emergy available decreases, which is reflected by cost escalation. Therefore, emergy in combination with cost can be considered as a criterion to evaluate the environmental sustainability of materials (Thomas and Praveen, 2020). Principles of econophysics can also be utilized in assessing the resource availability based on emergy. The fundamental law of statistical mechanics give the probability of states with different energy given by the Boltzmann–Gibbs exponential function expressed as

$$P(\varepsilon) = c \, e^{-\varepsilon/T} \tag{1}$$

where probability $P(\varepsilon)$ is the relative probability of a particular arrangement with given energy ε, T is the temperature, and c is the normalizing constant (Chakrabarti and De, 1997). Equation shows as energy increases the probability of an arrangement decreases. This process is analogous to emergy transactions in an ecosystem. So, to ensure sustainability in ecosystems, we need to limit the entropy of interactions. Boltzmann–Gibb's equation can be used to identify the changes in ecosystem entropy which captures the transition from rich state of ecosystem to its poorer state.

The most appropriate aggregates for construction are derived from riverine deposits, commonly occur in mountain and river valleys which are environmentally important. River sourced aggregates possess the advantages of being naturally sorted, can be easily extracted without the usage of specialized equipment and contain little organic matter. Extracting sediment storage and altering sediment transport, which occurs during sand mining, will enhance physical changes in the water channel, banks, bars, and floodplains. The localized sediment deposits support vegetation and habitats, which are utilized by riverine, riparian, and terrestrial species for spawning, feeding, and nesting (Owens et al. 2005). Deepening river bases also lowers the groundwater availability and aquifer recharge

The objectives of the research are (i) mapping of resource usage pattern and its impact on future availability and (ii) to predict the loss of ecosystem services based on the known consumptive practices of river sand. Therefore, the research presented in this paper establishes the utility of using fundamental equations of statistical mechanics to assess the state of ecosystem sustainability. The emergy of construction operations at regional level is mapped to selected time levels to elicit the time scale impact in the region based on the resource use pattern. Such representations made for a selected set of widely used materials are expected to map the point of vulnerability for consumptive pattern. The investigation further utilizes emergy analysis to quantify the lost environmental services, reflected due to inexpensive river sand extraction viz, groundwater recharge, nutrient building, and loss in ecosystem productivity.

2. METHODOLOGY

The state of Kerala, one of the smallest states in India which covers an area of 38,863 km² having a population of around 33 million is selected for this study. It borders the Arabian Sea to the west and the Western Ghats to the east having an undulated topography.

2.1. Accounting techniques on mapping of resource use pattern

The study applies logistic models to portray the trend of emergy available per unit cost of building materials across chosen time span. In the different models plotted, the asymmetric double sigmoidal function and Boltzmann function are employed to study the characteristics of the data obtained. UEV of building materials needed for analysis were gathered from existing literature. The market prices of building materials were collected from the Economics And Statistics Department, Government of Kerala. The UEVs used for calculations are based on the revised global emergy baseline (12 E+24 sej/year) GEB_{2016} (Brown et al., 2016). For evaluation, the average market price of building materials for 25 years were selected and energy available per unit cost values were computed for sand, gravel, brick, wood, and steel. Application of Gibbs–Boltzmann equation and logistic curves to derive environmental inferences of economic activity is discussed in the following sections.

2.2. Accounting of lost ecosystem services

2.2.1. Loss of nutrients

Emergy loss of nutrients calculated as

$$Em_{nu} = \sum (P_i * BD_i * A * D_i * UEV_{nu}) \tag{2}$$

Em_{nu} is the lost emergy of soil nutrients (sej); P_i is the percentage of i-th nutrient in soil (%); BD_i is the soil bulk density (g/m³); D_i is the soil depth (m); A is the area of soil strip (m²), UEV_{nu} is the specific emergy of i_{th} nutrient (sej/g).

2.2.2. Loss in groundwater recharge

Emergy loss in aquifers is computed by

$$Em_{gr} = \sum (H*S*A*D*G*UEV_{gr}) \tag{3}$$

Em_{gr} is the lost emergy of groundwater recharge (sej), H is the declined height in groundwater level (m), S is the specific yield (%), A is the area of assessment unit (m²), D is density of water (g/m³), G is the Gibbs free energy (J/g), UEV_{rs} is the transformity for groundwater (sej/J).

2.2.3. Decrease in channel natural storage

Emergy loss in aquifer storage capacity

$$Em_{gs} = \sum (P*A*H*D*G*UEV_{gs}) \tag{4}$$

where Em_{gs} is the lost emergy of groundwater natural storage (sej), P is the porosity (%), A represents the area of assessment unit (m²), H is the thickness of sand layer (m), D is water density (g/m³), G is the Gibbs free energy (J/g), and UEV_{gs} is the transformity for groundwater (sej/J).

2.2.4. NPP

Loss in NPP calculated as

$$Em_{NPP} = \sum (A * NPP_{sm}(avg) * UEV_{sm}) \tag{5}$$

Em_{NPP} indicates the lost emergy of NPP (sej); A is the area of assessment unit (m²); NPP_{sm} is the NPP of concerned ecosystem (g/m²/yr), UEV_{sm} means the UEV of concerned vegetation (sej/J).

2.2.5. Carbon sequestration

Loss in carbon sequestration assessed as

$$Em_{cs} = \sum (A * C_{avg} * UEV_{cs}) \tag{6}$$

$$UEV_{cs} = \frac{\sum MAX(R)}{NPP} \tag{7}$$

where Em_{cs} refers to emergy loss of carbon sequestration (sej/yr); A refers to the area of assessment unit (m²), C_{avg} is the average carbon sequestration of the concerned ecosystem (g/m²/yr), UEV_{cs} is the transformity of carbon sequestration (sej/g), $MAX(R)$ is the maximum value among renewable resources in the area (sej/m²/yr). NPP is the NPP of ecosystem (g/m²/yr).

3. RESULTS AND DISCUSSION

3.1. Variation in emergy available per unit cost across chosen time scales for selected materials

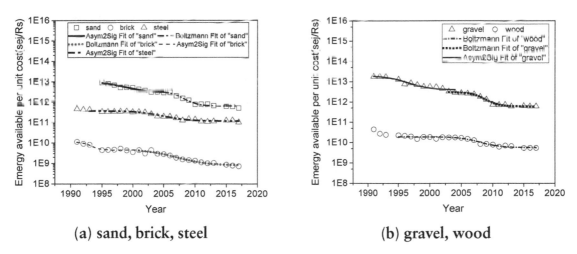

(a) sand, brick, steel   (b) gravel, wood

Figure 1. Plot of emergy available per unit cost across time for selected building materials. Data source: Economics and Statistics Department, Govt of Kerala.

Figure 1 shows the pattern of variation in available emergy per unit cost of selected building materials across years. Three classes of logistic material consumptive patterns were identified which currently exists in building material scenario: (a) material which has a combination of asymmetric double sigmoidal and Boltzmann pattern (sand, brick, and gravel), (b) materials which follows purely Boltzmann pattern (wood), (c) materials which exhibit only asymmetric double sigmoidal pattern(steel). In Figure 1(a), for sand, the asymmetric double sigmoidal function best suited the available data for early stages (1991–2005) and the Boltzmann function fitted for later stages (2005–2017). The fitted curve which highlights two distinct temporal relationships brings out the usage pattern of sand. Gentle slope at the top indicates, there was enough material for consumption till 2005, after that a steep decline in the availability of emergy per unit cost is observed. The decreasing trend in the emergy availability per unit cost of sand after 2005 is attributed to the scarcity in the resource supply. Compensated by the introduction of new substitute materials in the market and by intervention of rules and regulations for the extraction of resources, the system recovers and the pattern changes gradually. So, after 2012, the value in the Boltzmann function is decreasing at a very slow rate and gets stabilized at later stages. For gravel as shown in Figure 1(b). Asymmetric double sigmoidal function best captured the available data for years in the upper tail from 1991 till 2002 and the Boltzmann function for lower tail (2002–2017). Similar pattern is observed for brick as shown in Figure 1(a) with the asymmetric double sigmoidal fit up to 2002 and the Boltzmann fit for rest of span upto 2017. In the case of wood, as in Figure 1(b) it is observed that Boltzmann fit had started earlier around 1991 when compared to other natural materials like gravel and sand. Majority of construction sector was dependent on materials where the fuel source was wood. Since 1990, industrial roundwood is used tremendously by paper and packaging industry and construction sector. Rural small-scale brick making industries significantly used fuel wood, a major contributor to biomass-based energy. Utilization of wood directly or indirectly in the construction sector led to its exhaustion, one of the vital energy sources earlier than other natural resources. In the case of steel, as in Figure 1(a), only asymmetric sigmoidal fit is observed and Boltzmann situation is not initialized. Steel as an industrial material often does not undergo much price revisions over time resulting in stable pattern of emergy available per unit cost value.

3.2. Relevance of the models exhibited to determine variation in emergy available per unit cost

Sand, gravel, and brick, respectively, have a combination of asymmetric sigmoidal fit at earlier stages and Boltzmann fit at later stages. Sigmoidal curve shows the direction of material availability. Industrially procured materials as price revisions occur only due to slight changes in raw material costs, emergy available per unit cost value of steel moves in a stable pattern. In the models employed with the asymmetric double sigmoidal function and Boltzmann function, in the case of natural materials, it is found that there is a reduction in the emergy available per unit cost with time. Analogous to Boltzmann Gibbs' fundamental statistical theory, the total entropy of ecosystem increases, thereby the possibility of being in an ordered state decreases with time. Introduction of alternate mechanisms which limit the entropy of the system can decrease ecosystem disorderness thereby extending the formation of another Boltzmann scenario. In all the models, there seems to have a sudden fall which can be well represented by using Boltzmann function, and in the upper tail where slope is gentle when resources are in plenty asymmetric fit found more suitable. The tools of Gibbs–Boltzmann equation can be used to model the material availability which can be represented based on the emergy that exists in the material at different point of time. A systematic interpretation of ecosystems capacity to shift to alternative stable position with the absorption of industrial materials according to material demands can be drawn from displayed models.

3.3. Evaluation of the lost ecosystem services due to sand extraction

The emergy value of nutrients present in extracted sand is evaluated based on the geochemical parameters of bulk sediments of rivers in Kerala. The extraction of 1 m^3 of sand creates a nutrient loss value of about 1.29E+14, 4.03E+15, and 2.65E+15 $sej/m^3/yr$ for the respective phosphorous, iron, and organic carbon. The incision caused due to aggregate mining in rivers has led to lowering of water table in the adjacent areas. The emergy loss in environmental service due to declined recharge is 3.73E+09 $sej/m^2/yr$. The removal of sand from river bed results in the loss of aquifer storage space, which remains saturated throughout the year in an undisturbed condition. The emergy of the lost aquifer natural storage is about

4.03E+10sej/m³/yr for one cubic meter transfer of sand. The lowering of water level below the root zone leads to loss of riparian vegetation. The emergy loss in productivity per square meter of affected area was obtained as 8.04E+11 sej/m²/yr for riparian swamps and 2.09E+11 sej/m²/yr for riparian marshes. Carbon sequestration is a co-benefit of riparian vegetation. The emergy loss of carbon sequestration is 1.29E+14 sej/m²/yr for riparian swamps and 7.92E+11 sej/m²/yr for riparian marshes. Regulations imposed on sand extraction, limited the resource availability resulted in an increase in its procurement cost. The locally specific economic impact of this change is reflected in emergy available per unit cost of sand which shows a decreasing trend across years.

4. CONCLUSION

Construction sector on development economy directly depends on the availability of natural resources. Material usage pattern is mapped to assess its scarcity level at a regional scale which is directly related to cost fluctuations. An appropriate decision to ensure their supply in a sustainable way is necessary for regulatory bodies or government systems. In the present study, the time evolution of emergy available per unit cost in the region is modeled using sigmoidal curves based on data from economics and statistical department. There is an erosion in the emergy available per unit cost of natural materials with time. Based on that each and every resource passes through a rich period and a fall period. Prediction of that transition period from resource-abundant stage to resource-scarce stage is essential to ensure resource sustainability in the building industry. How availability of a material varies in a typical region in developing economy is considered and the significance of such an approach in the long-term planning is established. Economic losses for traded items are easily recognized by fluctuations of prices of commodities and services in markets. Emergy approach used in this research can be used to quantify non-economic losses and damages that occur for non-traded items in natural systems like damage to ecosystem services and biodiversity which is not yet efficiently addressed by current management approaches.

ACKNOWLEDGMENT

This work is supported by the AICTE (All India Council for Technical Education, India) through the QIP (Quality Improvement Programme).

REFERENCES

[1] Brown, M. T., and Ulgiati, S. "Energy quality, emergy, and transformity: H.T. Odum's contributions to quantifying and understanding systems," *Ecol Modell*, vol. 178, no. 1–2, pp. 201–213, 2004. doi: 10.1016/j.ecolmodel.2004.03.002.

[2] Brown, M. T., Campbell, D. E., de Vilbiss, C., and Ulgiati, S. "The geobiosphere emergy baseline: A synthesis," *Ecol Modell*, vol. 339, pp. 92–95, 2016. doi: 10.1016/j.ecolmodel.2016.03.018.

[3] Chakrabarti, C. G. and De, K. *Boltzmann Entropy: Generalization and Applications*, Kluwer Academic Publishers, 1997.

[4] G. Habert, Y. Bouzidi, C. Chen, and A. Jullien, "Development of a depletion indicator for natural resources used in concrete," *Resour Conserv Recycl*, vol. 54, no. 6, pp. 364–376, 2010. doi: 10.1016/j.resconrec.2009.09.002.

[5] Howard T. Odum, *Environmental Accounting : Emergy and Environmental Decision Making*. Wiley, 1996.

[6] Owens et al., "Fine-grained sediment in river systems: Environmental significance and management issues," *River Res Appl*, vol. 21, no. 7. pp. 693–717, 2005. doi: 10.1002/rra.878.

[7] Thomas, T. and Praveen, A. "Emergy parameters for ensuring sustainable use of building materials," *J Clean Prod*, vol. 276, 2020. doi: 10.1016/j.jclepro.2020.122382.

15. Estimation of Variation and Trend Analysis of Runoff in Karamana River Basin in Thiruvananthapuram

Akhila R* and Sindhu P

Department of Civil Engineering, College of Engineering Trivandrum

*Corresponding author: akhilak3a@gmail.com

ABSTRACT: Thiruvananthapuram is the densest district in Kerala according to the 2001 and 2011 censuses. There are many rivers flowing through this district out of which Karamana is a major river. It flows through densely populated portions and through regions containing several agricultural activities. It is also an important source of drinking water for Trivandrum city. Hence variation in its hydrology will affect a wide population. This study aims to analyze the variation characteristics and quantitively estimate the variation in runoff of the Karamana river basin over 40 years. The abrupt change points in the climatic parameters (temperature, rainfall) and runoff was identified using Mann–Kendall change point test and trend analysis of the parameters was carried out using Mann–Kendall test. Runoff in each of the change period and its variation was estimated.

KEYWORDS: Runoff, Trend Analysis, Mann–Kendall Test

1. INTRODUCTION

Climate change and its subsequent impacts are rising as a serious concern globally as it has brought about severe and permanent alterations to our ecosystem (Dad et al.,2021). These changes include the emergence of large-scale environmental hazards such as loss of biodiversity, increase in frequency of floods, increased snowmelt events, rise in sea water level, etc (Nuamah et al.,2019). Hence, understanding the pattern of climate change and its variability will be helpful for policymakers to plan accordingly and take suitable mitigation measures to reduce its impacts (Shao et al.,2021; Wu et al.,2020). From the past data, it is seen that Kerala usually experiences tropical monsoons with seasonally excessive rainfall and hot summer. However, recently Kerala is witnessing devastating incidents of climate change with rainfall being unevenly distributed throughout the year irrespective of the seasons and raised levels of temperature during the winter and monsoon seasons. Meteorological disasters such as floods are becoming an annual problem and severe flood occurred due to heavy rainfall during midyear of 2018 and 2019 in the state. Kerala also faced the threat of severe drought immediately after this.

Climate change such as change in temperature and rainfall patterns can affect runoff of a river basin in direct or indirect ways. The major impacts include the frequency of floods, drought events, and impacts on river flow and water quality. It also affects the volume and timing of water flow, affecting water resources, and the natural stability of the basins. Direct human activities like large-scale mining activities, water diversion, water withdrawal are the major factors leading to the runoff reduction in the strong impact period. The influence of land use changes like that on forest cover, agricultural land, grasslands, etc. also has a significant impact on runoff.

There are 44 rivers that flow through Kerala and about 50 major dams distributed mostly across the Western Ghats which provide water for agriculture and hydroelectric power generation. Flood occurrences in the years 2018 and 2019 resulted in overflowing of water along the banks of some of the rivers and most of the reservoirs in the state were near their full reservoir level (FRL) and most of the

Chapter 15 DOI- 10.1201/9781032657271-15

soil in the region became saturated which led to the opening of the shutters of almost all the major dams in Kerala. A combination of these torrential rains, overflow of rivers, and opening of the dam shutters resulted in severe flooding over the catchment areas. This has caused severe damage to the residents in the banks of the rivers, loss of habitats to the population around the banks, created a large number of relief camps, and has even led to many casualties.

The present study aims to estimate and analyze the variation of runoff in the Karamana river basin whose catchment area lies within the Thiruvananthapuram district. The main objective of the study is to quantify the variation in runoff in Karamana river basin over the last 40 years and carry out trend analysis of the same.

2. METHODOLOGY

2.1. Study area

The study area selected for the study was Karamana river basin that originates from the Western Ghats at Agastyar Koodam and flows through Trivandrum district in the state of Kerala. The river has a total length of approximately 68 km and a maximum elevation of 1717 m above mean sea level. The study area has an area of 605.88 km². The annual average rainfall in the river basin is about 2600 mm, average annual streamflow is 836 mm³, and the annual temperature ranges between 21 and 35°C.

2.2. Data collection

The required hydrometeorological data were obtained from Irrigation Design and Research Board, Trivandrum and Indian Meteorological Department, respectively, for a period of 40 years. The daily streamflow data is collected from the Mangattukadavu river gauge station and the meteorological data including daily maximum and minimum temperature and daily precipitation from two stations; Trivandrum airport and Trivandrum city. The methodology adopted for the study is shown in Figure 1.

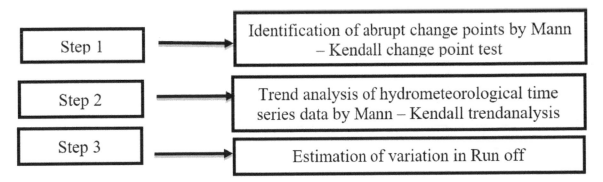

Figure 1. Methodology.

The identification of abrupt change points was carried out by Mann–Kendall change point test using the R Studio software. Statistic index UF of forward series, UB of reverse series, and Sequential Mann–Kendall plot were obtained as the output. The intersection point of UF and UB curves is considered as the change point if |UFi| ≤ UFα/2 within the significance level α of 5%. The change points in rainfall, minimum temperature, maximum temperature, and runoff for the study period 1980–2020 were identified and the period was divided into reference period and change period based on the change point. The trend analysis of runoff for each time periods within the change points was carried out using Mann–Kendall test. For testing the hypothesis, null hypothesis (H0) was that there is no trend in the data and alternative hypothesis (H1) was there is trend among the data, within the significance level of 5%. R Studio software was used for the analysis and the output parameters obtained include

Mann–Kendall statistic "Z," Sen's slope(s), and p value. For statistical significance, if the value of Z is greater than the standard normal deviation $Z_{1-\alpha/2}$, the null hypothesis is rejected and the trend is treated significant and vice versa. A positive value of test statistic "Z" is an indicator of an increasing trend and Sen's slope value gives the magnitude of a trend in a time series. After identifying the reference period, the mean annual runoff within the change points was estimated. The variation in runoff in each period with respect to the runoff in the reference period was then evaluated.

3.   RESULTS AND DISCUSSION

The abrupt change points in the time series data were identified using the Mann–Kendall change point test (Tables 1–7) and the variation in runoff in each change period was estimated (Table 11). Trend analysis was also carried out to understand the variation characteristics of the hydrometeorological parameters in each change period (Tables 8–10).

4.   MANN–KENDALL CHANGE POINT TEST

4.1. Rainfall

The abrupt change points in the time series data of rainfall from 1980 to 2020 at the meteorological stations were identified. The test results of the change point analysis of rainfall at Trivandrum airport are shown in Table 1 and of rainfall at Trivandrum city in Table 2.

Table 1. Test results of change point analysis of rainfall: Trivandrum airport.

Year	UF	UB	Critical Z value	Changepoint identification
1987	0.49	0.33	1.96	Significant change point
1988	0.21	0.49	1.96	Significant change point
1989	0.45	0.15	1.96	Significant change point
2016	1.05	1.32	1.96	Significant change point
2017	0.99	0.49	1.96	Significant change point

Table 2. Test results of change point analysis of rainfall: Trivandrum city.

Year	UF	UB	Critical Z value	Changepoint identification
1991	0.69	0.49	1.96	Significant change point
1995	1.08	1.10	1.96	Significant change point
1997	0.99	0.86	1.96	Significant change point
2000	1.21	1.38	1.96	Significant change point
2001	1.61	1.03	1.96	Significant change point
2002	1.24	1.49	1.96	Significant change point
2004	0.84	0.72	1.96	Significant change point
2009	1.62	2.07	1.96	Insignificant change point
2015	1.47	1.35	1.96	Significant change point
2016	0.97	2.07	1.96	Insignificant change point
2020	1.55	1.00	1.96	Significant change point

From the sequential Mann–Kendall plot, the years 1986, 1987, 1988, 1989, 2016, and 2017 were identified as the abrupt change points of rainfall in Trivandrum airport and the test results show that all the points are significant. Years 1991, 1995, 1997, 2000, 2001, 2002, 2004, 2009, 2015, 2016, and 2020 were identified to show an abrupt change in rainfall in Trivandrum city. All the years except 2009 and 2016 were found to be significant change points.

4.2. Maximum temperature

The abrupt change points in the time series data of maximum temperature from 1980 to 2020 at both the stations were identified and the test results are shown in Tables 3 and 4.

Table 3. Test results of change point analysis of maximum temperature: Trivandrum airport.

Year	UF	UB	Critical Z value	Changepoint identification
2014	1.52	0.99	1.96	Significant change point

Table 4. Test results of change point analysis of maximum temperature: Trivandrum city.

Year	UF	UB	Critical Z value	Changepoint identification
2001	2.88	2.20	1.96	Insignificant change point

The year 2014 was identified as an abrupt change point from the Sequential Mann–Kendall plot of maximum temperature at Trivandrum airport and it was identified as a significant change point within the significance level 5%. In Trivandrum city, the year 2001 was identified to show an abrupt change, however insignificant.

4.3. Minimum temperature

The abrupt changepoints in the time series data of minimum temperature from 1980 to 2020 at both the stations were identified and the test results are shown in Tables 5 and 6.

Table 5. Test results of change point analysis of minimum temperature: Trivandrum airport.

Year	UF	UB	Critical Z value	Changepoint identification
2000	2.11	2.00	1.96	Insignificant change point
2001	2.17	2.42	1.96	Insignificant change point
2003	1.89	1.57	1.96	Significant change point

Table 6. Test results of change point analysis of minimum temperature: Trivandrum city.

Year	UF	UB	Critical Z value	Changepoint identification
2014	1.01	0.25	1.96	Significant change point

The years 2000, 2001, and 2003 were identified as abrupt change points from the sequential Mann–Kendall plot of Trivandrum airport of which only the year 2003 was significant. In Trivandrum city, the year 2014 was the only change point identified and it was found to be significant.

4.4. Runoff

The change point analysis of runoff of the Karamana river basin was carried out for 40 years and the abrupt change points between the years 1980 and 2020 were identified. The test results of the change point analysis are shown in Table 7. The years 1992, 1999, 2003, 2008, and 2017 were identified as the abrupt change points with a significance level of 5%. Finally, all the results of the change point analysis of rainfall, maximum temperature, minimum temperature, and runoff were compared and the years 2003, 2008, and 2017 were identified as the common significant change points.

Table 7. Test results of change point analysis of annual runoff: Karamana river basin.

Year	UF	UB	Critical Z value	Changepoint identification
1992	1	0.15	1.96	Significant change point
1999	−0.21	0.54	1.96	Significant change point
2003	0.73	−0.04	1.96	Significant change point
2008	1.44	1.28	1.96	Significant change point
2017	0.77	0	1.96	Significant change point

4.6. Trend analysis using Mann–Kendall test

The trend analysis of annual rainfall, minimum temperature, and maximum temperature at both the meteorological stations Trivandrum city and Trivandrum airport was carried out and the results are tabulated as in Tables 8 and 9. An increasing trend was observed for all the parameters over the 40 years. The trend in annual rainfall showed a high Sens slope of 11.44.

Table 8. Mann–Kendall test results for whole period (1981–2020): Trivandrum city.

	Z	Sen slope	p value	Trend identification	Nature of variation
Annual rainfall	1.93	11.44	0.05	Significant trend	Increasing
Minimum temperature	3.41	0.02	0.0006	Significant trend	Increasing
Maximum temperature	6.36	0.03	0.68	Insignificant trend	Increasing

Table 9. Mann–Kendall test results for whole period (1981–2020): Trivandrum airport.

	Z	Sen slope	p value	Trend identification	Nature of variation
Annual rainfall	1.93	0.04	4.65E-05	Significant trend	Increasing
Minimum temperature	4.15	0.04	3.26E-05	Significant trend	Increasing
Maximum temperature	5.52	0.03	3.46E-08	Significant trend	Increasing

4.7. Trend analysis—runoff

The trend analysis of annual runoff in each time periods between the change points was carried out and the results are tabulated as shown in Table 10. Insignificant yet increasing trends were observed in the time periods 1981–2002 and 2003–2008 while decreasing trends were observed in the change periods 2009–2016 and 2017–2020.

Table 10. Mann–Kendall test results of annual trend analysis of runoff for each time period.

Year	Z-value	p value	Trend identification	Sen's slope	Nature of variation
1991–2002	0.94	0.35	Insignificant trend	12.86	Increasing
2003–2008	0.00	1.00	Insignificant trend	104.01	Increasing
2009–2016	0.00	1.00	Insignificant trend	−1.65	Decreasing
2017–2020	−1.02	0.31	Insignificant trend	−866.01	Decreasing

4.8. Estimation of variation in runoff

The variation in annual runoff of each change period with respect to the reference period was estimated and the results are as shown in Table 11. The runoff in the change period 2003–2008 has a variation of 2631 m³/s while the next period 2009–2016 has a decrease of 64.32 m³/s in runoff. The third change period 2017–2020 again shows an increase of 4425.25 m³/s.

Table 11. Variation in annual runoff in each change period with respect to the reference period.

Time period	Mean annual runoff (m³/s)	Runoff variation (m³/s)
1991–2002	420.05	
2003–2008	3051.05	2631.00
2009–2016	355.73	–64.32
2017–2020	4845.30	4425.25

5. CONCLUSIONS

The abrupt change points in the time series data were identified and trend analysis was carried out to understand the variation characteristics in each change period. The variation in runoff in each change period was then estimated quantitatively. The years 2003, 2008, and 2017 were identified as the abrupt change points in runoff over the last 40 years from 1980 to 2020. Trend analysis showed an increasing trend in runoff for the time periods 1980–2003 and 2004–2008. A decreasing trend was observed in runoff for the time periods 2009–2016 and 2017–2020. The runoff in the time period 2004–2008 was found to have increased to 3051.05 m³/s from a runoff of 420.05 m³/s in the reference period with a variation of 2631.00 m³/s. The runoff in the time period 2009–2016, however, got reduced to 355.73 m³/s with a decrease of 64.32 m³/s and that in period 2017–2020 got raised to 4845.30 m³/s with an increase of 4425.25 m³/s. Runoff variation may be impacted by climate change or human activities. Climatic parameters include variations in rainfall, temperature, etc. and human activities may include mining and quarrying, deforestation, and also significant land use changes in the basin with forest, cropland, and settlement. Hence mitigation measures are to be adopted suitably over the river basin so as to reduce the variations in its hydrology.

REFERENCES

[1] Awotwi, A., Anornu, G. K., Ballard, J. Q., Annor, T., and Forkuo, E. K., "Analysis of climate and anthropogenic impacts on runoff in the lower Pra River Basin of Ghana," *Heliyon*, vol. 3, pp. 2405–2440, 2017.

[2] Dad, J. M., Muslim, M., Rashid, I., Rashid, I., and Reshi, Z. A., "Time series analysis of climate variability and trends in Kashmir Himalaya," *Ecol Indic*, vol. 126, pp. 107690, 2021.

[3] Jin, S., Zheng, Z., Ning, L., "Separating variance in the runoff in Beijing's river system under climate change and human activities," *Phys Chem Earth*, vol. 123, pp. 103044, 2021.

[4] Liang, S., Wang, W., Zhang, D., Li, Y., and Wang, G., "Quantifying the impacts of climate change and human activities on runoff variation: Case study of the upstream of Minjiang River, China," *J Hydrol*, vol. 25, pp. 05020025, 2020.

[5] Nuamah, P. A., and Botchway, A., "Understanding climate variability and change: Analysis of temperature and rainfall across agroecological zones in Ghana," *Heliyon*, vol. 5, pp. 02654, 2019.

[6] Shao, Y., He, Y., Mu, X., Zhao, G., Gao, P., and Sun, W., "Contributions of climate change and human activities to runoff and sediment discharge reductions in the Jialing River, a main tributary of the upper Yangtze River, China," *Theor Appl Climatol*, vol. 145, pp. 1437–1450, 2021.

[7] Wang, S., Yan, Y., Yan, M., and Zhao, X., "Quantitative estimation of the impact of precipitation and human activities on runoff change of the Huangfuchuan River Basin," *J Geograph Sci,* vol. 22, pp. 906–918, 2012.

[8] Wu, L., Wang, S., Bai, X., Luo, W., Tian, Y., Zeng, C., Luo, G., and He, S., "Quantitative assessment of the impacts of climate change and human activities on runoff change in a typical karst watershed, SW China," *Sci Tot Environ,* vol. 601–602, pp. 1449–1465, 2017.

[9] Wu, L., Zhang, X., Hao, F., Wu, Y., Li, C., and Xu, Y., "Evaluating the contributions of climate change and human activities to runoff in typical semi-arid area, China," *J Hydrol,* vol. 590, pp. 125555, 2020.

16. Treatment of Commercial Kitchen Wastewater Using Protease Enzymes

Fazna Nazim[1,*] and Renu Pawels[2]

[1]Department of Civil Engineering, Cochin University of Science and Technology, Kalamassery

[2]Department of Civil Engineering, Cochin University of Science and Technology, Kalamassery

*Corresponding author: faznanazim@gmail.com

ABSTRACT: Kitchen wastewater, a type of grey wastewater has garnered significant attention because of organic pollutants, oil, and grease. The present water crisis demands the reuse of kitchen wastewater for nonpotable purposes. There are many kitchen wastewater treatment methods that are costly and time-consuming. So a cost-effective method is introduced for treating kitchen wastewater using enzymes. As enzymes are catalysts, they accelerate chemical reactions. These enzymes reaction is faster than the other techniques. Protease enzyme catalyzes proteolysis that breaks down proteins into small polypeptides. In this chapter, commercial kitchen wastewater (veg and non-veg) was treated with different quantities of protease enzymes and the optimum quantity was found at optimum pH and temperature conditions. From the results, it was found that 97% Biochemical oxygen demand (BOD) reduction (non-veg sample) using 0.3 g of enzymes after 5 days of treatment and 92% BOD reduction (veg sample) using 0.5 g of enzymes after 24 h of treatment.

KEYWORDS: Enzymes, Wastewater Treatment, Biochemical oxygen demand, Catalyzes, Kitchen Wastewater and Pollution

1. INTRODUCTION

Water is most important for the existence of all living forms. It is an easy solvent, enabling most pollutants to dissolve in it easily and contaminate it. Increasing population, urbanization, and industrialization have led to the deterioration of water. Water pollution is directly suffered by the organisms and vegetation that survive in water, including amphibians. Domestic and industrial waste is the most common cause of water pollution (Bansal et al., 2012). Kitchen waste is leftover organic matter, washing soap and detergent from restaurants, hotels, and households in which restaurants play a major role in discharging kitchen waste into the environment. In India, the restaurant industry is growing at a faster rate with wide range of cuisines and the diverse cooking techniques. Kitchen wastewater contains high organic, suspended solids, oil, and grease which cause harm to the environment and human health. Pollutants can also affect the ground waters.

Commercial kitchen wastewater is a type of grey wastewater that has garnered significant attention because of organic pollutants and oil and grease. It has hazardous effects on the environment, such as eutrophication in water bodies and choking of pipes in the sewer system. Stagnant Kitchen Waste Water invites serious diseases like malaria, filarial, and dengue. Reuse of Kitchen Waste Water in various fields helps to convert its zero value to a valuable resource. Further exploration of Kitchen Waste Water is required for determining its sustainable uses (Zdarta et al., 2021).

Enzymes are substances that act as a catalyst in living organisms, regulating the rate at which chemical reactions proceed without itself being altered in the process. The biological processes that occur within all living organisms are chemical reactions, and most are regulated by enzymes. Without enzymes, many of these reactions would not take place at a perceptible rate (Karn and Kumar, 2015). A large protein enzyme molecule is composed of one or more amino acid chains called polypeptide chains. The amino acid sequence determines the characteristic folding patterns of the protein's structure, which is essential to enzyme specificity. Bound to some enzymes is an additional chemical component called a cofactor, which is a direct participant in the catalytic event and thus is required for enzymatic activity. A cofactor may be either a coenzyme—an organic molecule, such as a vitamin—or an inorganic metal ion; some enzymes require both. A cofactor may be either tightly or loosely bound to the enzyme. If tightly connected, the cofactor is referred to as a prosthetic group (Varma et al., 2019).

Protease enzyme catalyzes the breakdown of proteins to shorter polypeptides or amino acids. They undergo proteolysis by hydrolyzing peptide bonds. In most of the living organisms, protease enzymes are essential for digestion and absorption of proteins. Proteases are found in all the living organisms, for example, bacteria, algae, plants, and animals and in some of the viruses too. Different types of protease enzymes remain active in the different pH ranges, for example, acid proteases, alkaline or basic proteases, and neutral proteases (Cammarota and Freire, 2006). Protease enzyme ranges from general to specific, for example, digestive protease enzyme, trypsin can cleave many proteins into smaller fragments, whereas enzymes like thrombin, which takes part in blood clotting are highly specific. Dietary protein is digested by many protease enzymes present in the digestive tract (Kad et al., 2020).

Protease enzyme is a group of proteolytic enzymes, which hydrolyze the peptide bonds present in proteins to convert it to shorter polypeptides and amino acids. They play a major role in the digestion, and absorption of dietary proteins.. In this paper, commercial kitchen wastewater(veg and non-veg) was treated with different quantities of protease enzymes, and the optimum quantity was found at optimum pH and temperature conditions (Javalkar et al., 2019).

2. MATERIALS AND METHODS

2.1. Sampling

The vegetarian and non-vegetarian samples used for the treatment were collected from Aryas restaurant and Malabar restaurant, Angamaly respectively. Sample was collected in 2 l of plastic bottle and was stored in refrigerator. Protease enzyme was bought online. pH of kitchen wastewater is acidic in nature. The protease enzyme activity observed to be maximum at optimum pH range was 8–10. Low pH suppresses the activity of the enzymes. The protease enzyme solution characteristics were found with time. Figure 1(a)–(c) represents non-vegetarian, vegetarian samples, and enzyme powder, respectively.

Figure 1. *(a) Non-vegetarian sample*   *(b) Vegetarian sample*   *(c) Protease enzyme powder.*

2.2. Characteristics of the commercial kitchen wastewater

The influent characteristics of the commercial kitchen wastewater are shown in Table 1.

Table 1. Characteristics of commercial kitchen wastewater.

Sl No.	Parameters	Unit	Non-vegetarian hotel Value	Vegetarian hotel Value
1	Color	Hazen	50	50
2	Odor		Objectionable	Objectionable
3	Turbidity	NTU	431	141
4	pH		4.8	4.27
5	Electrical conductivity	µS/cm	905	1109
6	Total suspended solids	mg/l	12000	6000
7	Biochemical oxygen demand	mg/l	4200	1200
8	Chemical oxygen demand	mg/l	12480	6400

2.3. Treatment of protease enzyme powder with vegetarian samples and non-vegetarian samples

The enzyme was added in different quantities, namely, 0.1, 0.3, 0.5, and 0.7 g, respectively, in 100 ml of vegetarian and non-vegetarian samples. The solutions were mixed thoroughly in a magnetic mixer.

Figure 2. (a) Treatment of protease enzyme powder in different quantities in vegetarian sample.

Figure 2. (b) Treatment of protease enzyme powder in different quantities in non-vegetarian sample.

Figure 2(a) and (b) shows the treatment of protease enzyme powder in different quantities viz., 0.1, 0.3, 0.5, and 0.7 g in 100 ml of vegetarian and non-vegetarian samples, respectively. The non-vegetarian samples were analyzed daily. Biochemical oxygen demand, Chemical oxygen demand, and Total suspended solids were checked daily. The general effluent standards are given in Figure 3.

3. RESULTS AND DISCUSSIONS

3.1. Optimization of protease enzyme quantity at optimum pH, temperature conditions—non-vegetarian samples

The factors considered during the optimization of an enzyme assay include the choice of buffer and its composition, the type of enzyme and its concentration, the type of substrate and concentrations, the reaction conditions, and the appropriate assay technology [8]. The optimum pH of commercial protease enzyme is in the range of 8–10. The samples were tested for Biochemical oxygen demand, Chemical oxygen demand, and Total suspended solids. The variation of Biochemical oxygen demand with respect to time is shown in Figure 7.

Parameters	unit	Inland surface water	Public sewer	Land for irrigation	Marine coastal areas
Biochemical oxygen demand (BOD)	mg/l	30	350	100	100
Chemical oxygen demand (COD)	mg/l	250	--	--	250
Total suspended solids (TSS)	mg/l	100	600	200	10-100

Figure 3. General effluent standards.

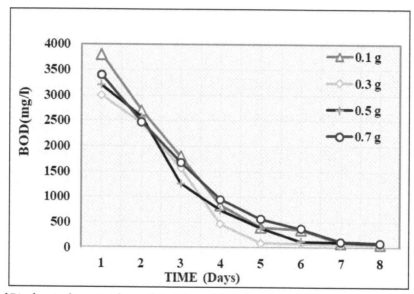

Figure 4. Variation of Biochemical oxygen demand in different quantities of protease enzyme in non-vegetarian sample.

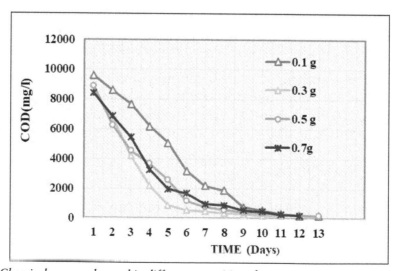

Figure 5. Variation of Chemical oxygen demand in different quantities of protease enzyme in non-vegetarian sample.

The variation of Chemical oxygen demand and Total suspended solids with respect to time is shown in Figures 4 and 5, respectively. When 0.1, 0.3, 0.5, and 0.7 g of protease enzyme were used to treat non-vegetarian sample, the Biochemical BOD value reduced to the limit after 7, 5, 8, and 8 days, respectively (Figure 7).

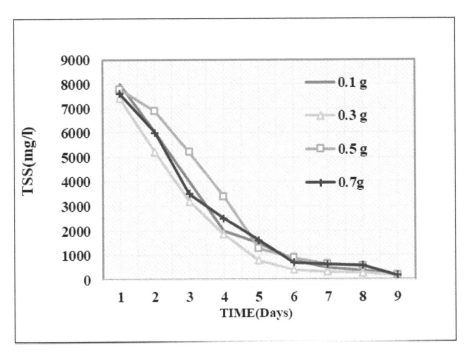

Figure 6. *Variation of Total suspended solids in different quantities of protease enzyme in non-vegetarian sample.*

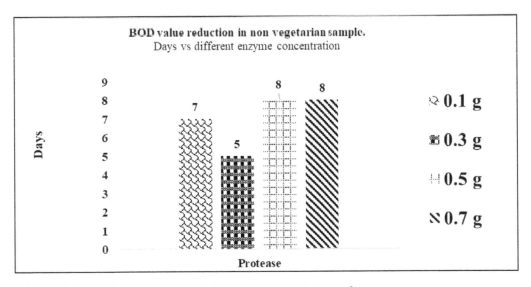

Figure 7. *Biochemical oxygen demand value reduction in non-vegetarian sample.*

3.2. Optimization of enzyme quantity at optimum pH and temperature conditions— vegetarian samples

The vegetarian samples were tested for Biochemical oxygen demand, Chemical oxygen demand, and Total suspended solids. The variation of Biochemical BOD with respect to time is shown in Figure 8.

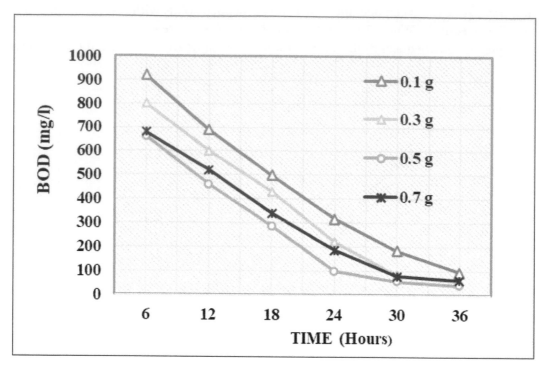

Figure 8. Variation of Biochemical oxygen demand in different quantities of protease enzyme in vegetarian sample.

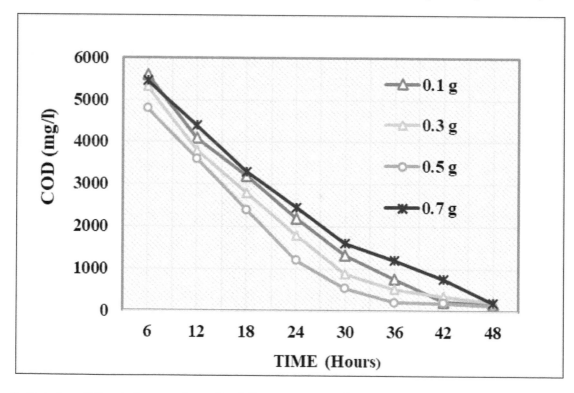

Figure 9. Variation of Chemical oxygen demand in different quantities of protease enzyme in vegetarian sample.

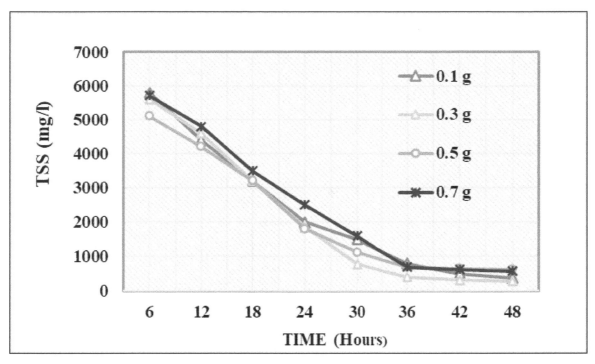

Figure 10. Variation of Total suspended solids in different quantities of protease enzyme in vegetarian sample.

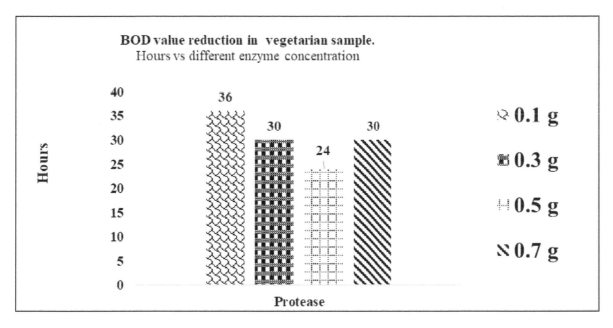

Figure 11. Biochemical oxygen demand value reduction in vegetarian sample.

When 0.1, 0.3, 0.5, and 0.7 g of protease enzyme were used to treat the vegetarian sample, the BOD value was reduced to the limit after 36, 30, 24, and 30 days, respectively (Figure 11).

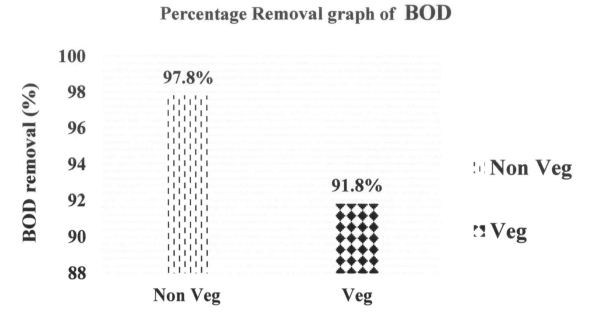

Figure 12. Percentage Biochemical oxygen demand reduction in non-vegetarian and vegetarian samples.

97.8% BOD was reduced while using 0.3 g of protease enzyme after 5 days of treatment and 91.8% BOD was reduced while using 0.5 g of protease enzyme after 24 hours of treatment. The treated effluent can be disposed to public sewers and can be used for land irrigation as per general effluent standards.

The analysis of non-vegetarian and vegetarian wastewater for protein content revealed a concentration of 183 and 1262 mg/l, respectively. The protein content was higher in vegetarian samples, hence the substrate concentration is more, and thus more faster the BOD reduction.

4. CONCLUSIONS

Waste management and treatment is one of the serious problem to the environment. About 50% of waste is organic in nature. If the discharge is not safe, it threatens both aquatic life and other living beings. Kitchen wastewater contains many highly toxic gases and chemicals. Presently, there are many methods for treating kitchen waste water but it is of high cost and time consuming [9].

As enzyme being a biocatalyst and need a substrate to function, it is used for treating kitchen wastewater. It is a less time-consuming technique and proteins present in kitchen wastewater acts as substrates for the enzymatic action. Normally enzyme reactions are used in solid wastes, but this technique was more useful in treating kitchen wastewater. Biochemical oxygen demand, Chemical oxygen demand, and turbidity of the wastewater was reduced to its permissible limit within a short time with the addition of protease enzyme.

Biochemical oxygen demand of non-vegetarian and vegetarian samples was reduced by 97.8% and 91.8%, respectively. It was found that enzyme treatment was a very effective and faster technique which can be used for the treatment of wastewater than other microbiological treatments. Hence it can be adopted widely as a safe and better disposal technique.

REFERENCES

[1] Bansal, N., Tewari, R., Soni, R., and Soni, S.K., "Production of cellulases from *Aspergillus niger* NS-2 in solid state fermentation on agricultural and kitchen waste residues," *Waste Manag*, vol. 7, pp. 1341–1346, 2012.

[2] Deepak Varma, Narain Anoop, Anoop Narain Singh, and A. K Shukla, "Use of garbage enzyme for treatment of wastewater," *Int J Sci Res*, vol. 7, pp. 7, 2019.

[3] Jakub, Zdarta, Katar Zyna, Jankowska, and Karolina Bachosz, "Enhanced wastewater treatment by immobilized enzymes," *Curr Pollut*, vol. 7, pp. 21, 2021.

[4] Javalkar Sayali D, Shinde Shruti C, Savalkar Shweta S, Pawar Sudarshan E, Dhamdhere Akash H, and Patil Shrikant, "Use of eco enzymes in domestic wastewater treatment," *Int J Inno Sci Res Technol*, vol. 2, pp. 2156–2465, 2019.

[5] Magali Christe Cammarota and Denise M. G. Freire, "Hydrolytic enzymes in the treatment of wastewater with high oil and grease content," *Bioresour Technol*, vol. 8, pp. 97, 2006.

[6] Nandini K Kad, Akshay R. Thorvat, and Naiem Harun Nadaf, "Experimental study on treatment of wastewater using garbage enzyme," *Int J Inno Res Sci Eng Technol*, vol. 6, pp. 5060–5066, 2020.

[7] Santosh Kumar Karn and Awanish Kumar, "Hydrolytic enzyme protease in sludge recovery and its application," *Biotechnol Bioprocess Eng*, 2015, doi: 15-0161-6

17. Analysis of Heavy Metal Adsorption Pattern in Soil Columns for Environmental Forensic Applications

Hemanth Manohar and George K Varghese*

Department of Civil Engineering, National Institute of Technology, Calicut, India

*Corresponding author: gkv@nit.ac.in

ABSTRACT: Pollution by heavy metals is a major concern because of its non-degradable nature, the potential for bio-magnification, and impacts on the exposed population. Identifying the source(s) of the heavy metals discharged on the soil is important to allocate site remediation responsibility and to take strict action against the polluters which can serve as deterrents to continued pollution. The duration of the discharge of pollutants to the soil is an important piece of information that will help in identifying the polluter. In this study, the pattern of adsorption of selected heavy metal pollutants on soil is used as an environmental forensic tool to estimate the duration of their discharge into the soil. Column studies were carried out on laterite soil by varying the concentrations and discharge duration of the heavy metals such as chromium (Cr) and copper (Cu) to obtain adsorption pattern of the metals on the soil. The adsorption pattern was characterized in terms of percentage saturation (heavy metals adsorbed/adsorption capacity of the soil for the particular metal) at different depths. The heavy metal concentrations in soil were determined using atomic absorption spectroscopy. Later, from the adsorption pattern, the discharge duration was estimated with reasonable accuracy. The method of predicting discharge history from adsorption patterns was found to be a reliable method with potential environmental forensic applications.

KEYWORDS: Environmental Forensics, Atomic Absorption Spectroscopy (AAS), Adsorption, Heavy Metal Contamination

1. INTRODUCTION

Contaminant transport is a complex phenomenon that is governed by a few physicochemical processes like advection-dispersion, diffusion, capillarity, etc., and some biological processes. Some of the major sources of contamination include pollution by industries, agricultural fields, landfills, septic tanks, leaky underground gas tanks, overuse of fertilizers and pesticides, etc. According to Robert D. Morrison (2015), environmental forensics is the use of systematic and scientific principles for the purpose of developing defensible scientific and legal conclusions regarding the source or age of a contaminant released into the environment.

Most of the source identification techniques employed before including methods like multivariate statistical analysis, GIS methods of approach, etc. (Amorosi et al., 2014; Dwivedi, 2014; Jianshu et al., 2014; Wei-Chih et al., 2016) that only consider the spatial variability of contaminants over a period of time and not vertical distribution. Grygar et al. (2016) and Rui Li et al. (2019) have analyzed vertical distribution of contaminants in river sediments. This study intends to do a similar approach as seen in sedimentary analysis, that is, analysis of vertical distribution of heavy metal contaminants in laterite soil.

This particular study aims to understand the absorption pattern of heavy metal in a soil column and the contaminant transport pattern in the saturated zone when subjected to flow of heavy metal-

Chapter 17 DOI- 10.1201/9781032657271-17

contaminated water. And hence use the above understanding to identify the time duration for which the contaminant was discharged by an industry or individual on a particular site which can be employed even decades after the contaminant discharge happened. The study will be carried out through laboratory simulations of transport in the vadose zone and saturated zone.

2. MATERIALS AND METHODS

2.1. Adsorbent

The adsorbent selected for this particular study is laterite soil procured from the National Institute of Technology Calicut, Kerala campus. The sample collected is dug out from 30 cm below ground level as the surface soil contains organic debris. The preliminary test on soil sample includes grain size analysis, pH analysis, and maximum dry density.

Table 1. Properties of soil.

Sl No.	Properties	Values
1	Optimum moisture content	10.5%
2	pH	5.7 @ 24.2°C
3	Maximum dry density	1.99 g/cm³

2.2. Adsorbate

A synthetic heavy metal solution was prepared by dissolving potassium dichromate ($K_2Cr_2O_7$) in deionized water to obtain a concentration of 10 ppm. The potassium dichromate powder was weighed on an electronic weighing balance with the least count of 0.0001 g.

2.3. Column adsorption studies

Column adsorption studies are conducted to find out, how the adsorption patterns are varying with time and depth in a continuous flow. The column layers were made of soil (10, 20, and 30 cm). Soil bed of different depths and 8.9 cm diameter was prepared and permeated first with water (blank solution) as trial. The chromium solution was introduced into the column at a flow rate of 60 mL min^{-1}. The influent solution concentration of chromium was 10 mgL^{-1}. The samples from the outlet were collected at different time intervals using sampling bottles. This process continued until the column gets exhausted. The amount of heavy metal in different samples was then analyzed for residual concentrations. After the column was exhausted, soil bed was taken out, divided into the number of layers, and analyzed for heavy metals.

3. RESULTS AND DISCUSSIONS

3.1. Column adsorption results of 10 cm adsorbent bed

The variations of outlet concentration for all the three cases have shown parabolic variations as in Figure 1.

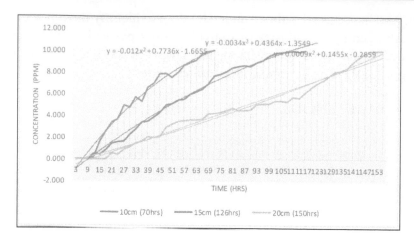

Figure 1. *Comparison of variation of outlet solution concentrations for all three (10, 15, 20cm) columns.*

Table 2. Variation of time elapsed for different percentages of adsorption of contaminant.

Time elapsed (h)	Percentage contaminant adsorbed (%)
22	40
23	39
26	42
43.5	73
65.67	100

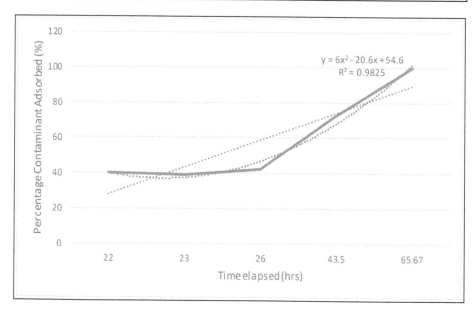

Figure 2. *Percentage contaminant adsorbed versus time graph.*

Figure 2 depicts the variation of percentage contaminant adsorbed on the soil with the time taken for it to do so. The curve has been smoothened and an equation generated with an R^2 value of 0.98. The generated equation is as follows,

$$y = 6x^2 - 20.6x + 54.6 \qquad (1)$$

The above equation gives the varying percentages of contaminant adsorbed on the soil with different times elapsed.

So, in a contaminated site, if we have obtained the quantity of contaminant adsorbed in soil, we can calculate how much percentage of the total capacity has the contaminant been occupied and in turn, we can calculate the time for which the contaminant flowed through the soil to reach that particular concentration.

This particular experiment has been done for an initial concentration of 10 ppm and lacks the data for other initial concentrations. To reach at a conclusion that the contamination was taken place at a particular point of time in the past, some corroborative information is needed which we are deficient of. However, attempts are done to compile data from previous studies to reach at conclusions showing the time elapsed for contamination of soil for varying initial concentrations of Cr(VI) which has been discussed below.

The above observations apply only for a contaminant with an initial concentration of 10 ppm. So, for generalizing the concept, attempts were made to establish a relation between a variety of initial contaminant concentrations and how in each case, the percentage removal varied. Two previous studies which dealt with adsorption studies were taken for the process of acquiring data on variation of percentage adsorption with varying initial concentrations. The following observations were generated by combining the data obtained from Mutongo (2014) and Syama (2015). Figure 3 shows the variation of adsorbed percentage of Cr(VI) quantity in a particular quantity of soil as the initial Cr(VI) concentration varies.

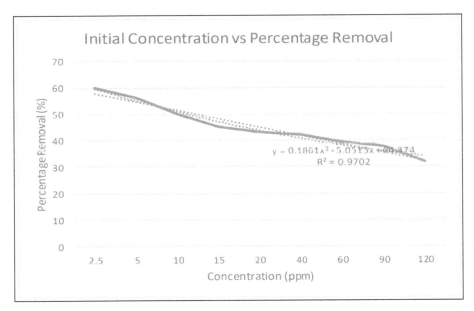

Figure 3. *Graphical representation of variation of percentage of contaminant adsorbed with varying initial contaminant concentration, as obtained from studies of Mutongo (2014) and Syama (2015).*

The equilibrium curve (Figure 3) shows that the overall percentage removal of Cr(VI) from the solution decreases with an increase in the initial Cr(VI) concentration. This may be attributed to the lack of sufficient surface area to accommodate much more metal available in the solution (Mutongo, 2014). Also, it was found that the amount of chromium removed from solution increases with an increase in concentration of Cr(VI). This was probably due to the higher interaction between the metal ions and metal sequestering sites of the adsorbent.

From the data obtained from the graph, divisors are generated in such a way that dividing the Cr(VI) quantity obtained in the field for an initial concentration other than 10 ppm (say 2.5 ppm) can be divided by the respective divisor to obtain the corresponding concentration as that would have been in the case of 10 ppm initial concentration.

The flow rates of the contaminant solution at the site under investigation can be found out from contaminant flow modeling.

Table 3. Divisors generated for dividing the in situ Cr(VI) quantities obtained for various initial Cr(VI) concentrations so as to be used in the same methodology as used for the 10 ppm method.

Initial contaminant concentration (ppm)	Divisor to be used to divide the in situ Cr(VI) quantity to convert in terms of 10 ppm method
2.5	0.833
5	0.893
10	1.000
15	1.111
20	1.163
40	1.190
60	1.266
90	1.323
120	1.572

4. CONCLUSIONS

The study was aimed at understanding the adsorption pattern of Cr(VI) in soil columns subjected to Cr(VI) contaminated water and using the understandings hence derived to predict the discharge history of heavy metals to laterite soils for environmental forensic applications. For this, column adsorption tests were carried out on three different soil columns of depths 10, 15, and 20 cm at an initial concentration of 10 ppm. The quantity of Cr(VI) adsorbed in the matrix of the soil in each case at varying time durations was determined and a relation was established between percentage contaminant adsorbed and time elapsed for the same. The relation obtained from the experiment was further back-tested to establish that it can be used to predict the time duration for which the contaminant flow occurred at a site under investigation.

The study proves helpful in environmental forensic applications where scientists/researchers can analyze the soil at the site under investigation for past contaminations and generate an estimate of the time duration for which the contaminant flow occurred. Although finding only the time duration is not enough to pinpoint the culprit, with certain corroborative data and evidence, the study can be improved and can zero down on the approximate time in the past when the contamination actually started and thereby reaching at a valid conclusion which will help them to find the culprit responsible, may that be an individual or industry and hence hold them accountable for the crime against environment and also for the remediation of the contaminated site within the specified time.

REFERENCES

[1] Amorosi, A., Guermandi, M., Marchi, N., and Sammartino, I., "Fingerprinting sedimentary and soil units by their natural metal contents: A new approach to assess metal contamination," *Sci Total Environ*, vol. 1, no. 500–501, pp. 361–372, 2014.

[2] Charlet, L., Manceau, A., "In situ characterization of heavy metal surface reactions: The chromium case," *Int J Environ Anal Chem*, vol. 46, no. 1–3, pp. 97–108, 1992.

[3] Debra M. Hausladen, Scott Fendorf, "Hexavalent chromium generation within naturally structured soils and sediments," *Environ Sci Technol*, vol. 51, no. 4, pp. 2058–2067, 2017.

[4] Diatta, J., and Kocialkowski, W., "Adsorption of zinc in some selected soils," *Polish J Environ Stud*, vol. 7, no. 4, pp. 195–200, 1997.

[5] Dube, A., Zbytniewski, R., Kowalkowski, T., Cukrowska, E., and Buszewski, B., "Adsorption and migration of heavy metals in soil," *Polish J Environ Stud*, vol. 10, no. 1, pp. 1–10, 2001.

[6] Ghosh, G. K., and Dash, N.,R., "Sulphate sorption–desorption characteristics of lateritic soils of West Bengal, India," *J Plant Animal Environ Sci*, vol. 2, no. 1, pp. 167–170, 2012.

[7] Kumar, N., and Mukherjee, I., "Effect of soil physicochemical properties on adsorption of tricyclazole," *J Agr Food Sci Tech*, vol. 4, no. 5, pp. 391–396, 2013.

[8] Lv, J., Liu, Y., Zhang, Z., and Dai, B., "Multivariate geostatistical analyses of heavy metals in soils: Spatial multi-scale variations in Wulian, Eastern China," *Ecotoxicol Environ Saf*, vol. 107, pp. 140–147, 2014.

[9] Maji, S. K., Pal, A., Pal, T., and Adak, A., "Adsorption thermodynamics of arsenic on laterite soil," *J Surf Sci Tech*, vol. 22, no. 3, pp. 161–176, 2017.

[10] Mertz, W., "Chromium: History and nutritional importance," *Biol Trace Elem Res*, vol. 32, no. 1–3, pp. 3–8, 1992.

[11] Osei, J., Simon, K. Y., Andrea, I. S., Faustina, A., and Francis, W. Y. M., "Impact of laterite characteristics on fluoride removal from water," *Gen Introd Chem Eng*.

[12] Oze, C., Bird, D. K., and Fendorf, S. "Genesis of hexavalent chromium from natural sources in soil and groundwater," *Proc Natl Acad Sci USA*, vol. 104, no. 16, pp. 6544–6549, 2007.

[13] Patil Mansing, R., and Raut, P. D., "Removal of phosphorus from sewage effluent by adsorption on laterite," *J Eng Res Tech*, vol. 2, no. 9, pp. 551–559, 2013.

[14] Robert D. Morrison, "Introduction to environmental forensics," In *Forensic Applications of Subsurface Contaminant Transport Models*, Chap. 16, pp. 555–591, 2015.

[15] Tembhurkar, A. R. and Dongre, S., "Studies on fluoride removal using adsorption process," *J Environ Sci Eng*, vol. 48, no. 3, pp. 151–156, 2006.

[16] Vogel, H.-J., Cousin, I., Ippisch, O., and Bastian, P., "The dominant role of structure for solute transport in soil: Experimental evidence and modelling of structure and transport in a field experiment," *Hydrol Earth Syst Sci*, vol. 10, pp. 495–506.

Theme: Geotechnical Engineering

18. Numerical Analysis of Diaphragm Wall Supporting Deep Excavation

Suruchi Sneha[1*], Syed Mohammad Abbas[1] and Sahu A K[2]

[1]Department of Civil Engineering, Jamia Millia Islamia, Jamia Nagar, New Delhi, India

[2]Department of Civil Engineering, Delhi Technological University, Shahbad Daulatpur, Main Bawana Road, Delhi, India

*Corresponding author: suruchi.sneha@gmail.com

ABSTRACT: The lack of available land is making higher skyscrapers and underground constructions increasingly necessary. The stability of the excavated regions as well as the stability of nearby structures is influenced by the deep excavations required for such structures in crowded places. To solve these problems, diaphragm walls, one of the soil-supporting structures, are being constructed. In this study, a diaphragm wall retaining a deep excavation of 14.2 m depth needed for three levels of integrated basement for an office complex in Noida was analyzed using PLAXIS 3D software. The outcomes were contrasted with measured data and 2D analytical results from past studies. This investigation found that the reported horizontal deflection of the wall and the estimated values from FEM analyses had a close similarity of horizontal displacement. Also, shear force and bending moment from 3D analysis were compared with that of 2D analysis performed in earlier research.

KEYWORDS: Deep Excavation, Diaphragm Wall, PLAXIS 3D, Horizontal Displacement, Shear Force and Bending Moment

1. INTRODUCTION

Because of the rise in population and also the lack of available land for building in major cities, civil engineers are needed to arrange for the vertical growth of cities in the form of skyscrapers and underground constructions. For the installation of these structures, deep excavation is highly needed. When extensive excavations are carried in densely populated areas where a structure is close by, the excavation causes severe settlement or tilting of the nearby structure as well as wall collapse in the excavated region. The solution to these issues is to use shoring systems like conventional retaining walls, secant pile walls, sheet pile walls, and diaphragm walls to hold deep excavation. Although a number of researchers have worked on various deep excavation support system components, there are not many case studies that have been published on these systems, particularly when it comes to diaphragm walls. Only a small number of them did studies 3D analyses on diaphragm wall. In light of this, the purpose of the current study is to do a 3D analysis of the diaphragm wall sustaining a deep excavation at a selected site and then verify the findings by contrasting the predicted and measured values. Data from inclinometers installed at the site and 2D analysis values were obtained for this purpose from a case study conducted by Nisha and Muttharam (2017).

2. LITERATURE REVIEW

Many studies have been conducted in the past to examine the installation method of diaphragm walls as well as their impact on the stability of the excavated area. In Europe, USA, and Asia, studies

Chapter 18 DOI- 10.1201/9781032657271-18

and reports on the impacts of retaining structure installation on the response of the neighboring ground have been made for various soil types. These include research projects by Davies, Tedd et al. (1984), Tamano et al. (1996), Farmer and Attewall (1973), Dibiagio and Myrvoll (1972), and Farmer and Myrvoll. Their investigations covered topics like the impact of installing different kinds of earth retaining structures on ground motions and alterations in horizontal earth pressure and pore pressure, as well as numerical modeling of the consequences of wall installation. According to Chan and Yap (1992), the installation of diaphragm wall panels had an impact on the performance of the nearby old masonry structures on soft or loose soil. As indicated in the reference, numerous other researchers have simulated the diaphragm wall adopting various methods of analysis to estimate the lateral displacement of the diaphragm wall supporting deep excavations (Conti et al., 2012; Gourvenec and Powrie, 1999; Li et al., 2021; Masuda et al., 1994; Merritt et al., 2010; Poh et al., 2001). These research projects used different finite element software to model the diaphragm wall in three dimensions. A comparison of the outcomes of 2D and 3D analyses of raft and piled raft foundations was conducted by Balakumar et al. (2018), and it was found that while 2D analyses can produce results with reasonable accuracy for the design office's needs during the initial trials, thorough 3D finite element studies should be used for the final draft. Important findings from a study by Nam and Dung (2020) using Plaxis 3D software include the correlation between the SPT number and the vertical Young's modulus of soils, as well as observations about how the deformation of the diaphragm wall is influenced by the unload-reload Young's modulus and how the surrounding soils are displaced vertically as a result of the vertical Young's modulus of soils.

Thus, Plaxis 3D has only been employed by a few number of researchers, particularly when analyzing diaphragm walls. Hence, the objective of this research is the modeling and analysis of diaphragm wall and validation of the model with the measured data.

3. SITE DESCRIPTIONS

For the present study, the location of the site has been chosen as Noida (India). Details on the location were gathered from a prior case study by Nisha and Muttharam (2017). It is observed that ground water table is 6 m below the normal ground level and the soil profile is mainly composed of two kinds of soil: silty sand with a little bit of gravel every now and then, and sand with a little bit of silt at a deeper level. The area had a topographic profile that was essentially flat and a ground elevation of 200.58 m.

3.1. Input soil parameters and interface

Table 1 displays the soil properties and interface taken into account for PLAXIS analysis. From the prior case study by Nisha and Muttharam (2017), the characteristics of the soil profile were discovered.

Table 1. Input soil parameters and interface used in 3D Plaxis analysis (Data from Nisha and Muttharam, 2017).

Layers (m)		Type of soil	Unit weight of soil (kN/m3)	Saturated unit weight of soil (kN/m3)	Young's modulus (kN/m2)	Poissons ratio, μ	Friction angle, φ (degrees)
Top	Bottom						
0	4	Silty sand (SM)	15	16	22,900	0.25	27
4	8	Sand (S)	16	17	35,900	0.25	29
8	11	Sand (S)	18	19	47,500	0.25	29
11	17	Sand (S)	19	20	64,400	0.25	30
17	22	Sand (S)	19	20	78,400	0.35	30

22	28	Sand (S)	19	20	80,000	0.35	30
>28		Sand (S)	20	21	85,000	0.35	31
Material model		Mohr–Coulomb					
Material type		Drained					
Strength reduction factor		0.67					

3.2. Input parameters of diaphragm wall and soil anchors

It was suggested to put soil anchors at three levels to maintain the diaphragm wall while the basement was being excavated. From the prior case study, all the information regarding the diaphragm wall and the design of the soil anchors was gathered (Nisha and Muttharam, 2017). Maximum depth of excavation and embedment depth of D-wall was 14.2 and 7 m, respectively. For the D-wall of 800 mm thickness, three layers of anchors (3, 7, and 11 m from finished road level) were recommended. For each layer, the horizontal distance between soil anchors was 1.66 m c/c. Comprehensive design and calculations for soil anchors and length of anchor with failure plane were performed in the previous case study (Nisha and Muttharam, 2017).

4. 3D FINITE ELEMENT ANALYSIS OF D-WALL

Diaphragm wall modeling process using Plaxis 3D software consisted of three phases. Phase 1 was considered to be the soil's initial condition. The Mohr–Coulomb model was used to simulate soil behavior. Phase 2 involved the modeling of a diaphragm wall as well as modeling of anchors. The diaphragm wall of 21.2 m depth was simulated as plate element, free length, and fixed length of soil anchors were input as node-to-node anchors and geo-grid elements, respectively. Phase 3 of the model simulates an excavation of 14.2 m depth by deactivation of soil layers up to that depth behind the D-wall.

5. 3D ANALYSIS OUTCOMES AND DISCUSSION

4.1. Horizontal deflection

From the analysis, it was observed that the highest and lowest horizontal displacements were 0.0277777 and 0 m, respectively. Figure 1 illustrates the variation in horizontal displacement with diaphragm wall depth. For the purpose of showing a comparative graph, the values of the real horizontal displacement and the anticipated values from the 2D analysis from a prior study by Nisha and Muttharam (2017) are employed. After the completion of the excavation activity, the inclinometers E-79, N-52, and N-21 reported the maximum deformation of approximately 7, 15, and 27 mm, respectively. The inclinometer N-21 showed greater observed displacements. The horizontal displacement curves for diaphragm walls of 800 mm thickness following the final level of excavation are compared for different depth of diaphragm wall in Figure 2.

According to the curves drawn, the wall's numerical deformation behavior and the actual deformation behavior as measured by inclinometers have a comparable profile. However, the movements that were actually realized were consequently less than the values discovered through analysis. This could be due to a number of factors, such as uneven soil, expected design surcharges that the walls did not really suffer on the site, and more. Moreover, 3D analysis yields lower values for horizontal deflection than 2D analysis does. This demonstrates that 3D analysis yields data that are more reliable than 2D analysis.

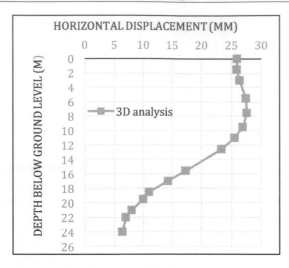

Figure 1. *Total horizontal displacement versus diaphragm wall depth graph.*

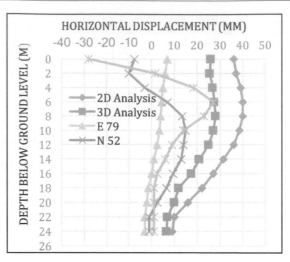

Figure 2. *Graph comparing the results of a case study and 3D finite element analysis.w.*

4.2. Shear force

The shear force curve derived from 3D analysis is displayed in Figure 3. Following 3D analysis, the maximum and minimum shear forces on the diaphragm wall are 220.7 and –155 kN/m, respectively, whereas they were 186 and –143 kN/m, respectively, following 2D analysis carried out in a prior study (Nisha and Muttharam, 2017). Figure 4 contrasts the shear force on the diaphragm wall between 3D and 2D analyses. The graph demonstrates that the design maximum and minimum shear forces determined from the 2D and 3D analyses are almost comparable.

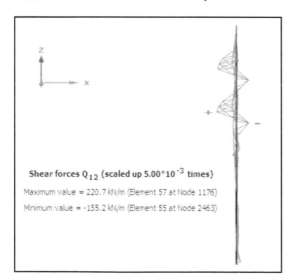

Figure 3. *Shear force diagram generated from 3D Plaxis as output.*

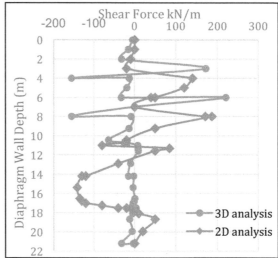

Figure 4. *Shear force diagram with depth comparison plot from 3D analysis and 2D analysis (Data from Nisha and Muttharam 2017).*

4.3. Bending moment

The bending moment graph from 3D analysis can be seen in Figure 5. After 3D analysis, the maximum and minimum bending moments of the diaphragm wall are 115.2 and –442 kNm, respectively. In contrast, 2D analysis from earlier study (Nisha and Muttharam, 2017) yielded values of 106.3 and –503.5 kNm.

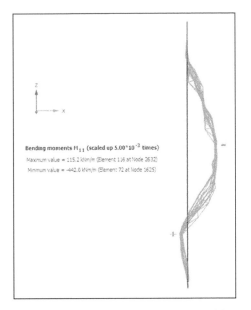

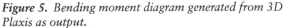

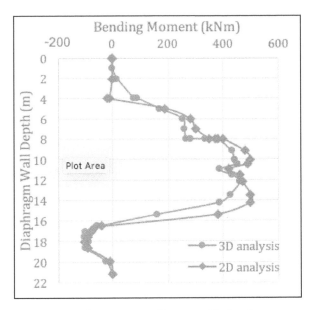

Figure 5. *Bending moment diagram generated from 3D Plaxis as output.*

Figure 6. *Bending moment diagram with depth comparison plot from 3D analysis and 2D analysis (Data from Nisha and Muttharam 2017).*

Figure 6 compares the diaphragm wall's bending moment in cases of 3D and 2D studies. It is clear from the graph that the design maximum and minimum bending moments determined by the 2D and 3D analyses are almost similar.

CONCLUSION

The investigation of the behavior of the diaphragm wall holding the underground basement was the main goal of this effort. The diaphragm wall having 800 mm thickness underwent 3D study, and after the excavation was finished, its behavior was observed. The findings were compared to in situ measurements and 2 D analysis findings from earlier research [1]. Here is a list of the conclusions drawn from monitoring data and the projected diaphragm wall response. As a result of the 3D analysis, the diaphragm wall's maximum and minimum lateral deflections are shown to be 0.027777 and 0 m, respectively.

a) It was discovered, after contrasting the 3D analysis results with the field measurements and 2D analysis results, that the expected behavior of diaphragm wall follows a pattern that is similar to that of the behavior observed and 2D analysis.

b) According to inclinometer records, 2D and 3D analytical outcomes, the highest recorded horizontal wall deflection ranged from 25 to 90% of the 3D analytical data, which was larger than that of the 2D analytical data (17.5–67.5%).

c) It is also clear from the outcome data that the horizontal deformation values from the 3D analysis fall between the real values and the values from the 2D analysis. As a result, it may be said that the values obtained through 3D modeling are more reliable than those obtained through 2D modeling.

d) The maximum shear force derived from the 3D assessment varied by 18.3% of the results calculated from the 2D assessment. While the minimum shear force derived from 3D model was 8.4% of that discovered through 2D model.

e) The highest bending moment's percentage fluctuation from 3D parametric study was 8.5% of the value from 2D parametric study. While the minimum bending moment determined by 3D parametric study only varies by 12% of the value determined by 2D parametric study.

f) The 3D model of the diaphragm wall is verified and may be employed for further study since the values produced from the 3D analysis have a similar profile to the values that were seen during the site inquiry. Comparing the three-dimensional and two-dimensional outcomes also reveals that 3D modeling is effective in obtaining more precise results. Thus, this research offers helpful information for using 3D modeling to analyze the behavior of diaphragm walls.

ACKNOWLEDGMENTS

The authors express gratitude to Department of Civil Engineering, Jamia Millia Islamia, Jamia Nagar, New Delhi, India for their active contribution in this study.

REFERENCES

[1] Balakumar, V., Min Huang, Erwin Oh, and Balasubramaniam, A. S., "A critical and comparative study on 2D and 3D analyses of raft and piled raft foundations," *Geot Eng*, vol. 49, no. 1, pp. 150–164, 2018.

[2] Chan Sin F., and Teck F. Yap, *Effects of Construction of a Diaphragm Wall Very Close to a Masonry Building*. ASTM International, 1992.

[3] Conti Riccardo, Luca de Sanctis, and Giulia Viggiani, "Numerical modelling of installation effects for diaphragm walls in sand," *Acta Geotechnica*, vol. 7, no. 3, pp. 219–237, 2012.

[4] Dibiagio, E., and Myrvoll, F., "Full scale field test of a slurry trench excavation in soft clay," Fifth Eur Conf On Soil Proc/Sp/, no. Conf Paper. 1972.

[5] Farmer, I. W., and Attewell, P. B., "Ground movements caused by a betonite-supported excavation in London Clay," *Geotechnique*, vol. 23, no. 4, pp. 576–581, 1973.

[6] Gourvenec, S. M., and Powrie, W., "Three-dimensional finite-element analysis of diaphragm wall installation," *Geotechnique*, vol. 49, no. 6, pp. 801–823, 1999.

[7] Jasmine Nisha, J., and Muttharam, M., "Deep excavation supported by diaphragm wall: A case study," *Indian Geote J*, vol. 47, no. 3, pp. 373–383, 2017.

[8] Li Hu, Ruchun Wei, Henghua Zhu, Chao Jia, Hao Shang, Xin Wang, Shuang Li, and Guoping Shi. "Numerical analysis of excavation stability of transfer station of Jinan Metro," *IOP Conf Series: Earth Environ Sci*, vol. 632, no. 2, p. 022019. IOP Publishing, 2021.

[9] Masuda Toru, Herbert H. Einstein, and Toshiyuki Mitachi, "Prediction of lateral deflection of diaphragm wall in deep excavations," *Doboku Gakkai Ronbunshu*, vol. 1994, no. 505, pp. 19–29, 1994.

[10] Merritt, A. S., Menkiti, C. O., Harris, D. I., Zdravkovic, L., Potts, D. M., and Mair, R. J., "3D finite element analysis of a diaphragm wall excavation with sacrificial crosswalls," Geotechnical Challenges in Megacities, GeoMos 2010, Int. Geotech. Conf., Moscow, Russia. 2010.

[11] Nam Nguyen Hong, and Nhu Van Dung, "Effect of deep excavation on deformation of diaphragm wall and adjacent structures," In *Geotechnics for Sustainable Infrastructure Development*, pp. 337–344. Singapore: Springer, 2020.

[12] Poh Teoh Yaw, Anthony Teck Chee Goh, and IngHieng Wong, "Ground movements associated with wall construction: Case histories," *J Geot Geoenviron Eng*, vol. 127, no. 12, pp. 1061–1069, 2001.

[13] Tamano Tomio, Satoshi Fukui, Hiromasa Suzuki, and Kano Ueshita. "Stability of slurry trench excavation in soft clay," *Soils Found*, vol. 36, no. 2, pp. 101–110, 1996.

[14] Tedd, P., Chard, B. M., Charles, J. A., and Symons, I. F., "Behaviour of a propped embedded retaining wall in stiff clay at Bell Common Tunnel," *Géotechnique*, vol. 34, no. 4, pp. 513–532, 1984.

19. Soil Stabilization of Clayey Soil With Mix of Stone Dust and Groundnut Shell Ash

Pawandeep Singh and Ajay Vikram
Department of Civil Engineering, Rayat-Bahra University, Mohali, Punjab
Corresponding author: pawandeeg11@gmail.com, ajayvikram99151@gmail.com

ABSTRACT: Since soil is created as a result of the weathering and disintegration of rocks, it is the least expensive and most easily accessible material for structural engineering. Engineering structures are constructed on top of this stratum of the earth's crust. But because different soil types include different amounts of minerals, the soil characteristics change from place to place, Due to the mineral's montmorillonite, illite, and kaolinite, clayey soil exhibits the greatest shrinkage and swelling. It is possible that the soil on a construction site would not have the qualities necessary to satisfy the design specifications, making it unsafe to build on these kinds of soil strata. In this case, it is not economically feasible to import the right soil from distant sources to replace the existing soil strata, so here soil stabilization plays an important role. soil stabilization is a technique used to enhance the stability of unstable soils in order to accomplish engineering objectives. in this study, materials used for soil stabilization are stone dust with groundnut shell ash (GSA). Stone dust is obtained during crushing of stones and it is a waste material, groundnut shell is an agriculture waste and ash from groundnut shell is categorized under pozzolana, as it is seen in the testing and research that on mixing these with soil it shows higher shear strength, increases MDD, decreases OMC, etc. To observe and investigate the variations in soil properties caused by these additives, several tests such as California bearing ratio, unconfined compressive strength, standard Proctor test, and others are carried out on soil samples with the addition of materials in varied amounts.

KEYWORDS: Soil Stabilization, Stone Dust, Groundnut Shell Ash, California Bearing Ratio, Compaction

1. INTRODUCTION

Soil stabilization enhances the properties of soil to withstand stresses in every weather condition. Materials used in this research work are a mix of stone dust and groundnut shell ash (GSA). Different tests performed on soil with replacement in certain proportions of these materials and values are noted. Test performed is the Standard Proctor Test (SPT) used to determine bearing capacity and settlement of soil, unconfined compression strength (UCS) is used to determine maximum axial compressive stress that a soil sample can bear, California bearing ratio (CBR).

However, no one has ever used stone dust with the addition of groundnut shell ash. Other researchers have used stone dust alone or with the inclusion of specific materials like lime, cement, rice husk ash, stabilizers, etc., which can often be expensive and show gaps in study. Due to their availability and the fact that they are waste materials, these materials are relatively inexpensive to purchase. Therefore, by conducting this research, we are improving soil stability and reducing waste.

Stone Dust and GSA are sieved through BS 200 No.75µ aperture sieve before addition to clayey soil.

When soil is replaced with stone dust in proportions of 0%, 5%, 10%, 15%, 20%, and 25%, the optimum values are noted at 20% of stone dust. Next, we take a soil sample with 20% of the optimum amount of stone dust and carry out all of the tests by adding GSA in proportions of 0%, 2%, 4%, 6%,

Chapter 19 DOI- 10.1201/9781032657271-19

8%, and 10%, at which point the appropriate proportion is noted. Following the completion of all testing, we now see the optimum ratios for combining the two materials.

2. MATERIAL USED

2.1. Clay

Because clay is an expansive soil, civil engineers often run into issues when designing and building structures. Clay soil swells as it comes into touch with water and shrinks when the amount of water in the soil drops, both of which cause serious structural damage. These swelling and shrinkage properties can be overcome by soil stabilization.

The source of soil used for this study is from Villages—Mehta, Bathinda (151001), Punjab Table 1 shows the properties of clay soil.

Table 1. Properties of clay soil.

Properties	Value
Water content (%)	26.8
Specific gravity	2.71
Liquid limit (%)	56
Plastic limit (%)	29.67
Plastic index (%)	26.33
OMC (%)	24
MDD (kN/m3)	15.1
UCS (kN/m2)	119
CBR	4.78

2.2. Groundnut shell ash

Groundnut shell obtained from the local shopkeeper in Bathinda city. Groundnut shell was ashed in open atmosphere and is then passed through 75 μ aperture sieve Table 2 shows the chemical composition of groundnut Ash.

Table 2. Chemical composition of GSA.

Sio2	34.81
Al2O3	7.74
Fe2O3	2.17
Cao	10.8
Mgo	4.16
Na2O + K2O	24.28
TiO2	1.12
Fe2O3	2.69
SO3	6.41
CO3	6.2

2.3. Stone dust

Stone dust is created when rocks are crushed in a machine to manufacture crushed stone. Depending on the kind of stone that was put through the machine, its specific composition will vary.

Stone dust used in this research is bought from Nalagarh Stone Crusher Kurali Road, Mohali Sector 67, Chandigarh 160062 Table 3 shows the chemical composition of Stone Dust.

Table 3. Chemical Composition of stone dust.

Sio2	65.71
Al2O3	19.32
Fe2O3	5.26
Cao	3.61
Mgo	2.19
K2O	2.26
TiO2	1.27

3. TEST RESULTS AND DISCUSSION

All these tests are conducted on virgin soil samples first with only stone dust and then mix of GSA with optimum value of SD (20%) in proportions of 2%, 4%, 6%, 8%, and 10% of soil sample.

3.1. Standard proctor test

SPT is used to determine the soil compaction properties like optimum moisture content at which soil can achieve maximum dry density Figure 1. shows the variation in MDD and OMC values with different mix of GSA % in soil with 20% stone dust as given in Table 4.

Table 4. Test Results of MDD and OMC at varying percentage of GSA with optimum value of SD (20%)

Soil (%)	Stone dust (optimum value)	% of GSA	Maximum dry density (kN/m3)	Optimum moisture content
100	0	0	15.1	24
80	20	2	16.69	20.8
78	20	4	16.98	20
76	20	6	17.51	19.2
74	20	8	17.73	18.55
72	20	10	16.76	18.59

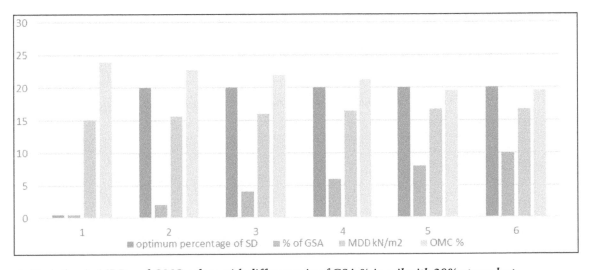

Figure 1. Variation in MDD and OMC values with different mix of GSA % in soil with 20% stone dust.

3.2. CBR

CBR is used to measure the strength of the soil with a standardized penetration test Figure 2. shows the variation in CBR values with different percentage of stone dust as given in Table 5. and Figure 3. shows the variation in CBR values with different mix of GSA in soil with 20% stone dust as given in Table 6..

Table 4. Variation in MDD and OMC values with different mix of GSA % in soil with 20% stone dust as shown in Table 4.

Soil + SD %	Soaked CBR value (2.5mm)
Virgin soil	4.78
5%	9.2
10%	16.1
15%	20.4
20%	23.9
25%	23.2

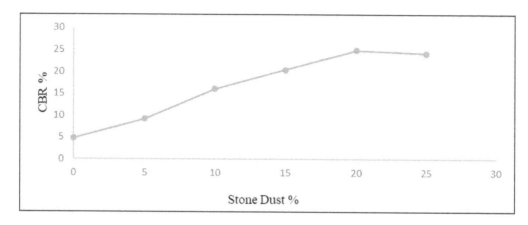

Figure 2. Variation in CBR with different percentage of stone dust.

Table 5. Test results of CBR for different % of GSA with optimum value of stone dust.

GSA % + optimum SD %	Soaked CBR value (2.5mm)
Virgin Soil	4.78
Soil+ 2% GSA+ 20% SD	18.87
Soil+ 4% GSA+ 20% SD	15.35
Soil+ 6% GSA+ 20% SD	17.68
Soil+ 8% GSA+ 20% SD	21.74
Soil+ 10% GSA+ 20% SD	22.16

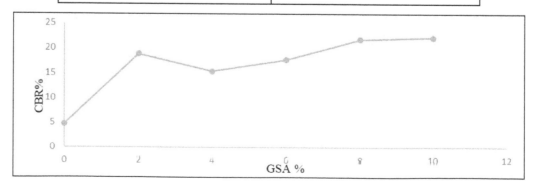

Figure 3. Variation in CBR values with different mix of GSA in soil with 20% stone dust.

3.3. UCS

Unconfined compressive strength (UCS) is used to calculate the highest axial compressive stress that a specimen may tolerate.

Table 6. Test results of UCS for different % of GSA with optimum value of stone dust.

Soil + SD %	UCS (kN/m²)
Virgin soil	119
5%	128
10%	136
15%	143
20%	151
25%	142

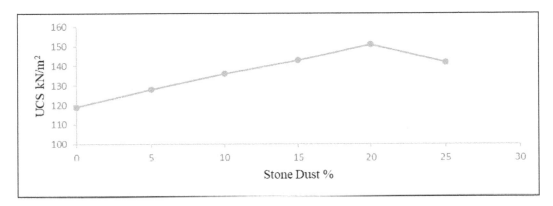

Figure 4. Variation of UCS with different % of stone dust.

Table 7. Test results of UCS for different % of GSA with optimum value of stone dust.

GSA % + Optimum SD %	UCS (kN/m2)
Soil+ 2% GSA+ 20% SD	149
Soil+ 4% GSA+ 20% SD	128
Soil+ 6% GSA+ 20% SD	154
Soil+ 8% GSA+ 20% SD	215
Soil+ 10% GSA+ 20% SD	223

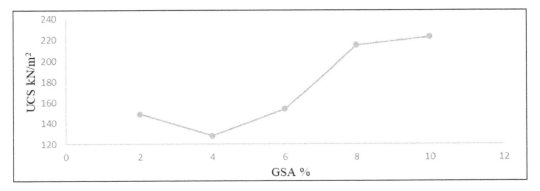

Figure 5. Variation in UCS values with different mix of GSA in soil with 20% stone dust.

Table 8. All test results.

Stone dust (%)	GSA (%)	OMC (%)	MDD (kN/m3)	UCS (kN/m2)	CBR
0	-	24	15.1	119	4.78
5	-	23.2	15.5	128	9.2
10	-	21.16	15.8	136	16.1
15	-	21.2	16.2	143	20.4
20	-	19.9	16.6	151	23.9
25	-	19.1	16.4	142	23.2
20	2	20.8	16.69	149	18.87
20	4	20	16.98	128	15.35
20	6	19.2	17.51	154	17.68
20	8	18.55	17.73	215	21.74
20	10	18.59	16.76	223	22.16

able 9 shows the test results of GSA, OMC, MDD, UCS, CBR with different proportions of SD and GSA mix together as shown.

4. CONCLUSION

From all the results following conclusions can be drawn:

- Optimum percentage of stone dust is 20% and optimum percentage of GSA is 8%. As further increase of these materials leads to decrease or slight increase in values of UCS and CBR.
- On addition of 20% SD and 8% GSA, UCS values increased by 175%. It shows a greater increase in the compressive strength of soil.
- CBR gives a value of 4.78–21.74, improving subgrade strength and properties.
- CBR values decreased at 20% SD + 8% GSA as compared to only 20% SD because of a change in shear strength of soil due to the addition of another material.

REFERENCES

[1] Haldar, M. K., and Das, S. K., "Effect of sand stone dust for quartz and clay in triaxialporcelain composition," *Indian Acad Sci*, vol. 35, no. 5, pp. 897–904, 2012.

[2] Hunter D., "Lime induced heave in sulphate bearing clay soils," *ASCE J Geot Eng* vol. 114, pp. 150–167, 1988.

[3] Ijimdiyaa, T. S., Ashimiyu, A. L., and Abubakar, D. K., "Stabilization of black cotton soil using groundnut shell ash," *Elect J Geotech Eng (EJGE)*, vol. 7, no.4, pp. 3645–3652, 2012.

[4] Modak, P. R., Nangare, P. B., Nagrale, S. D., Nalawade, R. D., and Chavhan, V. S., "Stabilization of black cotton soil using admixtures," *Int J Eng Innov Technol*, vol. 1, no. 5, pp. 11–13, 2012.

[5] Nadgouda, K. A., and Hegde, R. A., "The effect of lime stabilization on properties of black cotton soil," Proceeding Indian Geotechnical Conference-2010 (GEOtrendz), Bombay, pp. 511–514, 2010.

20. Studies on the Performance of Under-reamed Piles in Clayey Sand Through Model Tests

Ajai Joy Nariyelil [1,*], Minu J [1], Benny Mathews Abraham [1,2], Sobha Cyrus[1] and Deepa Balakrishnan S [1]

[1] School of Engineering, Cochin University of Science and Technology, Kochi, India

[2] Albertian Institute of Science & Technology, Kochi, India

*Corresponding author: ajaijoynariyelil@gmail.com

ABSTRACT: Pile capacity estimation and prediction of their behavior in different soil conditions are critical in construction industry. Even though many model studies were conducted to study the behavior of under-reamed pile in different soil conditions, none of them came across their behavior in clayey sand. Therefore, this paper discusses an experimental investigation that aims to understand various aspects, such as bulb angle, bulb spacing and number of bulbs, impacting the capacity of under-reamed piles in clayey sand. Nine model piles in total were tested, including a normal pile, four single under-reamed piles with different bulb angles, and four double under-reamed piles with different bulb spacings. Load-settlement curves are obtained from the model tests, and various criteria reported in the literature are used to determine the capacity of these piles. The pile capacity increased with the increase in number of bulbs. The pile with bulb angle of 45° and with bulb spacing of $1.5D_u$ (D_u is the bulb diameter) showed maximum capacity for single and double under-reamed piles, respectively. The test results provided a clear indication of the kind of under-reamed piles that should be installed at construction sites that contain clayey sand.

KEYWORDS: Bulb Angle, Bulb Spacing, Under-Reamed Piles

1. INTRODUCTION

If some weak strata exist on top of a robust stratum, pile foundations are usually employed to effectively transmit weight from the superstructure to deeper levels. Under-reamed pile foundations are one of such deep foundations that is employed when the underlying soil is not strong enough. They might be cast in place piles or bored compaction piles. In India, under-reamed piles have been used extensively since the 1950s. The biggest benefit of under-reamed piles over other types of piles is considered to be the under-reams because acceptable performance can be attained with lower stem diameters and lengths. This suggests that during the design phase, consideration is also given to the economic characteristics of under-reamed piles.

Cast-in-place concrete piles with one or more bulbs along their stems are referred to as under-reamed piles, and they have a greater capacity to withstand compression and tension loads (Farokhi et al., 2014). The ultimate load of the piles could be worked out using time-settlement plot from the pile load test results (Nazir et al., 2015). These piles can increase tip resistance and pile shaft friction while reducing the effects of negative skin friction (Chandrasekaran et al., 1978). Under-reamed pile failure in soil masses is significantly influenced by the shapes of the large bulbs in the pile. The under-reamed piles' geometrical design may influence how the soil interacts with the piles and the sort of failure that occurs to the soil mass around them (Hamed et al., 2019). According to the studies conducted by Pakrashi (2017) on numerous under-reamed piles of varying diameter and length at various sites,

Chapter 20 DOI- 10.1201/9781032657271-20

the safe pile capacity as determined by subsoil parameters differs from the safe pile capacity given in the IS 2911-Part 3. The conical shape of the bulb is favored over other under-reamed bulb shapes in terms of reducing vertical displacement, according to the findings of numerical analysis by Jebur et al. (2020) on the geometrical form of reams and their final bearing capacity under dynamic load. According to Jebur and Ahmed (2020), these bulbs improve the friction, bearing surface, and the pile capacity in compressive stress. The number, shape, size, and location of the bulbs had a major impact on the bearing capacity of the piles in compression and tension (Kurian and Srilakshmi, 2010). Vali et al. (2019) found that there was no significant increase in the bearing capacity with an increase in overall pile length but it increased with an increase in the number of bulbs. A number of studies were also carried out by Christopher and Gopinath (2016), Shetty et al. (2015), and Shrivastava and Bhatia (2008) to comprehend the impact of the geometrical characteristics of the under-reamed pile on its ultimate capacity.

From the literature, it is evident that numerous research works have been done in sand and clay, to understand the compressive behavior of under-reamed piles in these soil. This study compared the performance of single and double under-reamed piles installed in clayey sand with that of a normal uniform diameter pile. The most effective pile out of single under-reamed piles with various bulb angles and double under-reamed piles with various bulb spacings was identified through analysis.

2. MATERIALS

2.1. Soil

Clayey sand used for the experimental study was prepared by thoroughly mixing the marine clay collected from an under-reamed piling site and the properly sieved river sand which was available in the laboratory in a definite proportion. The water content of the soil was maintained throughout the experimental study. The properties of the clayey sand used for the experiment is given in Table 1.

Table 1. Properties of Clayey Sand used in the study.

Properties		Values
Moisture content (%)		41
Bulk unit weight (kN/m³)		17.75
Angle of internal friction (degrees)		13
Cohesion (kPa)		4.0
Particle size distribution	Sand size (>0.075 mm) (%)	60
	Silt and clay size (<0.075 mm) (%)	40

2.2. Under-reamed Pile Models

Timber is used to construct the under-reamed pile models(Mahogany). There are a total of nine distinct pile models produced. Four of them are single under-reamed piles with various bulb angles of 15°, 30°, 45°, and 60°. Then, there are four double under-reamed piles with different bulb spacings:$1.0D_u$, $1.25D_u$, $1.5D_u$, and $1.75D_u$. The final pile created is a normal pile which has a stem diameter of 4 cm, which is the same as the diameter of the under-reamed piles. For both single and double under-reamed piles, the bulb diameter is maintained constant at 100 mm. All nine piles are maintained at 450 mm in length. The timber pile models used in the study are shown in Figures 1 and 2.

Figure 1. Models of normal and single under-reamed piles.

Figure 2. Models of double under-reamed piles.

2.3. Metal tank

The pile model tests are conducted in a metal tank with measurements of 450 × 450 × 600 mm whose size was set based on the pile diameter. Boundary effects are thought to be insignificant considering the size of the tank.

3. METHODOLOGY

3.1. Filling the tank

Clayey sand was filled in the tank in layers of 100 mm thickness (six layers). It was filled by a procedure combining the method of filling clay and sand. The clayey sand was thrown into the tank as in the case of clay and then each layer was compacted using the cylindrical weight to ensure maximum density. As in the case of the other soil types, after filling 150 mm from the bottom of the tank, the model pile to be tested was placed and then the remaining of the tank was filled following a similar procedure. The top was leveled once the tank was filled and experimental set-up was established.

3.2. Loading mechanism

A hydraulic loading mechanism is used for this study. By pumping oil, load is applied, and a proving ring with a capacity of 10 kN is used to measure the load. The tank is placed on the loading frame and load is applied using a hydraulic jack.

3.3. Test procedure

The test tank is filled with clayey sand using the processes mentioned in the section above. The external compressive load is gradually introduced using the hydraulic jack. Proving ring attached with the test setup is used to measure the external load that has been applied. Two dial gauges are used to measure the settlement. Until the displacement reaches 12 mm or the proving ring starts to give a stable reading with continuous settling, load is applied. Finally, using the load-settlement curves plotted from the collected data, the ultimate capacities corresponding to 12 mm, 7.5 mm (7.5% of D_u), and 4 mm (10% of D, i.e., the stem diameter) settlements are determined.

4. RESULTS AND DISCUSSIONS

4.1. Load tests on single under-reamed pile

the load-settlement curve for under-reamed piles with varying bulb angles is shown in Figure 3 and from the figure, it is quite evident that the capacity is maximum for the pile with 45° bulb angle. Single under-reamed pile with bulb angle 45° has clearly the maximum ultimate capacity corresponding to all the settlements considered and the minimum capacity was obtained for the pile with 60° bulb angle even though its capacity was comparable to that of the pile with 15° bulb angle. The maximum capacity observed for 45° bulb angle agrees with the suggestions of IS 2911-Part 3.

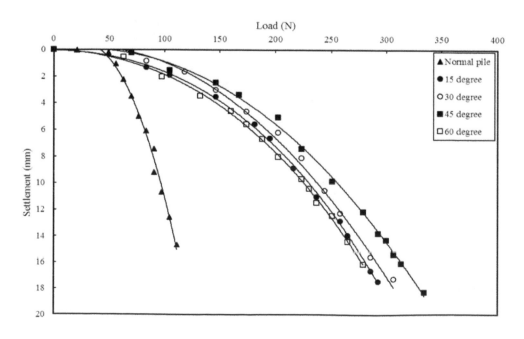

Figure 3. *Load-settlement curves for single under-reamed piles tested in clayey sand.*

4.2. Load tests on double under-reamed piles

Load-settlement curves as given in Figure 4 show that the double under-reamed pile with a bulb spacing of $1.5D_u$ has a slightly higher capacity as compared to others. The capacity of the pile with $1.75D_u$ bulb spacing has the least capacity as compared to all other double under-reamed piles. The maximum capacities of the piles with respect to bulb spacings increased in the order of $1.75D_u$, $1.0D_u$, $1.25D_u$, and $1.5D_u$.

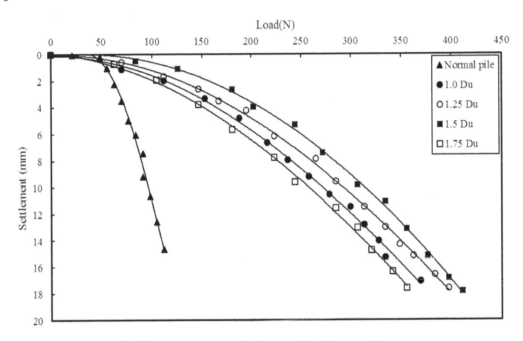

Figure 4. *Load-settlement curves for double under-reamed piles tested in clayey sand.*

4.3. Comparison of capacities of piles with increase in number of bulbs

The maximum capacities of normal piles, single under-reamed piles (with 45° bulb angle) and double under-reamed piles are given in Table 2. From this data, it is evident that as the number of under-reams or bulbs increases the capacity of the pile also increases. Here, the introduction of a single bulb alone increases the capacity of a normal pile by 2.8 times and for a double under-reamed pile it increases further to almost 3.5 times as that of a normal pile.

Table 2. Capacities of piles corresponding to 12 mm settlement.

Pile	Capacity (N)
Normal pile	100
Single under-reamed pile	280
Double under-reamed pile	345

5. CONCLUSIONS

The goal of the current study was to understand different parameters, such as bulb angle, bulb spacing, and number of bulbs, that affect the capacity of under-reamed piles in clayey sand. It was established through a series of model studies that the aforementioned factors truly influence the capacity of under-reamed piles. The following conclusions were drawn from the study:

- For a single under-reamed pile, the maximum capacity was observed in the pile with a bulb angle of 45°.
- In the case of double under-reamed piles, the maximum capacity was observed for the pile with bulb spacing of $1.5D_u$.
- The capacity increased by almost 180% when a bulb was introduced to normal uniform diameter pile.
- The capacity increased by almost 20% when the number of bulbs of the under-reamed piles increased from one to two.

The acquired results agree with the recommendation of IS 2911-Part 3 to provide an under-ream angle of 45°, which is generally followed on site. The findings also recommend to provide a bulb spacing of $1.5D_u$ for double under-reamed piles in site containing clayey sand.

REFERENCES

[1] Chandrasekaran, V., Garg, K. G., and Prakash, C., "Behaviour of isolated bored enlarged base pile under sustained vertical loads," *Soils and Foundations*, vol. 18, no. 2, pp. 1–15, 1978.

[2] Christopher, T., and Gopinath, M. B., "Parametric study of under-reamed piles in sand," *International Journal of Engineering Research & Technology (IJERT)*, vol. 5, no. 7, pp. 577–581, 2016.

[3] Farokhi, A. S., Alielahi, H., and Mardani, Z., "Optimizing the performance of under-reamed piles in clay using numerical method," *Electronic Journal of Geotechnical Engineering*, vol. 19(Bundle G), pp. 1507–1520, 2014.

[4] Hamed, M., Canakci, H., and Khaleel, O., "Performance of multi-helix pile embedded in organic soil under pull-out load," *Transportation Infrastructure Geotechnology*, vol. 6, no. 1, pp. 56–66, 2019.

[5] Jebur, M. M., Ahmed, M. D., and Karkush, M. O., "Numerical analysis of under-reamed pile subjected to dynamic loading in sandy soil," *In IOP Conference Series: Materials Science and Engineering, IOP Publishing*, vol. 671, no. 1, p. 012084, 2020.

[6] Jebur, M. M., and Ahmed, M. D., "Experimental investigation of under reamed pile subjected to dynamic loading in Sandy soil". *In IOP conference series: materials science and engineering) IOP Publishing*, vol. 901, no. 1, p. 012003, 2020.

[7] Kurian, N. P., and Srilakshmi, G., "Studies on the geometrical features of under-reamed piles by the finite element method," *Journal of Karunya University*, vol. 2, no. 1, pp. 1–14, 2010.

[8] Nazir, R., Moayedi, H., Mosallanezhad, M., and Tourtiz, A., "Appraisal of reliable skin friction variation in a bored pile," *Proceedings of the Institution of Civil Engineers-Geotechnical Engineering*, vol. 168, no. 1, pp. 75–86, 2015.

[9] Pakrashi, S., "A comparative study on safe pile capacity as shown in Table 1 of IS 2911 (Part III): 1980". *Journal of The Institution of Engineers (India): Series A*, vol. 98, no. 1, pp. 185–199, 2017.

[10] Shetty, P., Naveen, B. S., and Kumar, N. B. S., "Analytical study on geometrical of underreamed pile by Ansys," *International Journal of Modern Chemistry and Applied Science*, vol. 2, no. 3, pp. 174–180, 2015.

[11] Shrivastava, N., and Bhatia, N., "Ultimate bearing capacity of under-reamed pile-finite element approach," *In The 12th International Conference of International Association for Computer Methods and Advances in Geomechanics (IACMAG)*, pp. 1–6, 2008.

[12] Vali, R., Khotbehsara, E. M., Saberian, M., Li, J., Mehrinejad, M., and Jahandari, S., "A three-dimensional numerical comparison of bearing capacity and settlement of tapered and under-reamed piles," *International Journal of Geotechnical Engineering*, vol. 13, no. 3, pp. 236–248, 2019.

21. Characterization and Stabilization of Problematic Clayey Soil at Maniyaparambu, Kottayam

Sangeetha J, Sankaran V, Diya A T and Rejoice A A*
Department of Civil Engineering, Rajiv Gandhi Institute of Technology, APJ Abdul Kalam Technological University, Kottayam, India
*Corresponding author: rejoiceabraham@rit.ac.in

ABSTRACT: Clay soil, typically expansive clays is among the most problematic soils, posing a challenge to civil and geotechnical engineers. It has numerous problems due to high moisture content, low shear strength, and high compressibility. Volumetric changes due to changes in moisture regime are troublesome and put repeated stress and hence serious damage to the structure. Therefore, it is absolutely necessary to characterize the clay soil on site before any construction activities are carried out on it. In this study, attempts have been made to critically analyze the different properties of clay soil collected from a site near Maniyaparambu, Kottayam. The area is constituted entirely by clayey soil rich in organic content. Also, it is characterized by high moisture content, low strength, and high plasticity. In view of this soil condition, an attempt was made to stabilize the soil sample. The use of magnesium chloride as a stabilizer for clayey soils has been investigated in terms of its effects on the liquid limit, plastic limit, and vane shear strength parameters of treated clay. The optimum percentage of Magnesium chloride was determined based on the improvement in properties achieved.

KEYWORDS: Stabilization, Magnesium Chloride, Atterberg Limits, Vane Shear Strength

1. INTRODUCTION

Clay is a natural material consisting primarily of fine-grained minerals. It is composed of tiny particles possessing plastic and adhesive properties. Clay also contains small voids and pores, and so it is capable of retaining large quantities of water. In this condition, it tends to expand and shrink, that may lead to settlement (Ahmad, 2014). When subjected to increments in water content, clay tends to soften and liquefy. With its low strength and stiffness, clay often causes difficulties in construction. This has challenged geotechnical engineers because weak soil may cause damage to the foundation of buildings and poor subgrade results in developing cracks along the road pavement (Ahmad, 2014). The presence of clayey soil at a construction site may not be suitable for the desired purpose, being a highly complex and variable material. However, due to the rapid growth of population, it is impossible to avoid constructing on clayey soils. Hence it is absolutely necessary to try out some methods of improvement of these soils.

Soil stabilization is a technique introduced many years ago with a main aim of rendering deficient soils capable of meeting the requirements of specific engineering projects. It could be done by several methods like mechanical, chemical, or by soil inclusions. Cementitious materials are several binding materials that may mix with water to form a plastic paste and thereby improving several of their properties. Hence, stabilization of soil using cementitious material is an option to solve this inherent problem of clay soil (Makusa, 2012). The area of study, Cheeppunkal–Maniyaparambu road prominently consists of clay soil that is well known for its poor engineering properties. In addition to this, the area was badly affected by the massive floods of August 2018 further worsening the condition.

Groundwater table rose to a shallow depth causing instability of the subgrade soil. Also, a large amount of settlement was visible during the preliminary examination, focusing on the need for analyzing its properties and finding out a suitable method of stabilization. The study focuses on the characterization and stabilization of the soil sample collected from a site at Maniyaparambu. The objective of the study is to determine a suitable stabilizer for the particular soil, to study the effect of stabilization and to determine its optimum percentage.

2. LITERATURE REVIEW

Clay is an aggregate of microscopic and sub-microscopic particles derived from the chemical decomposition of rock constituents. It is a very fine-grained, cohesive, and plastic soil consisting of hydrated aluminum silicates and the size of clay particles falls below 2 μm. In clay, a soil grain usually consists of only one mineral and the electrical forces acting at the surface of the clay particle is predominant than gravitational force due to its small size. In clay, the particles are always in contact with water, and the properties of soil depend on the interactions between the clay particles, the water, and the various dissolved materials in the water (Terzaghi et al., 2010).

2.1. Clay–water interaction

The study on clay–water interaction is equally important as its knowledge about composition and structure because it greatly affects the cohesion and plasticity characteristics of the soil. The charge on the surface of clay is negative. The intensity of the charge and the mineralogical composition of the particle are the reasons for the activity of clay. The positive end (H+) of the water molecule will attach to the negative surface charge of clay as water molecule is dipolar. In the close vicinity of clay surface, the adsorbed water will behave like a solid. As the distance from the surface increases, the property of water changes to highly viscous liquid. Beyond the influence zone, there is no attraction on the water and is called free water while the water in the influence zone is called adsorbed water. To drive off the water in the immediate vicinity of the particle surface, the soil has to be treated to a temperature higher than 200°C (Arora, 2015; Murthy, 1995; Ural, 2018). The very large adsorbed layer of water is the reason for the ability of soil to deform plastically without cracking when various amounts of water are added to it. This plasticity character is obtained as the particle moves over one another with the support of very large viscous interlayers. The cohesion of particles is imparted by the shear strength of adsorbed layers and not due to interparticle attraction. The addition of salt will result in the base exchange and hence the formation of the adsorption complex. The change formed in the soil water by the addition of salt will manipulate the plasticity character of soil (Ayyar, 2000). Due to the intrinsic problem-causing properties of clayey soil, before constructing any structures on this type of soil, a better understanding of the problems caused by them is necessary.

2.2. Stabilization of clay soil

If the soil available for construction works is not suitable for their intended purpose, engineers go for alternatives like stabilization of soil which is a ground improvement technique. Soil stabilization is the process of improving the engineering properties of the soil so as to improve its strength and durability. The stabilization techniques are usually mechanical or chemical stabilization. In mechanical stabilization, the properties of soil are improved by changing its gradation and by compacting the mixture to the desired density. The mechanical stability of the mixed soil depends upon mechanical strength of the aggregate, mineral composition, gradation, plasticity characteristics, and compaction. This method is commonly used for improving the low-bearing strength of subgrade (Arora, 2015). Whereas in chemical stabilization, suitable additives are added to the soil for improving its properties and the additives used are cement, fly-ash, lime, or a combination of these and chloride compounds. These additives alter the physical and chemical properties of the soil including the cementation of the soil particles (Onyelowe and Chibuzor, 2012). Clayey soils are found all over the world. Since it poses a great challenge to any construction happening on it, engineers from all over the world have done immense research and is still continuing studies on clayey soils. Researchers have focused on studying its properties and behavior with altering water content to propose a suitable method for its stabilization.

3. METHODOLOGY OF THE STUDY

The methodology for the study consists of three parts: reconnaissance and preliminary exploration, basic characterization of the soil sample, and stabilization

3.1. Reconnaissance and preliminary exploration

Maniyaparambu is located in Arpookkara panchayat of Kottayam district. The site is a paddy field, surrounded by streams on three sides. The construction of Cheeppunkal–Maniyaparmbu road was in progress. The first reach was completed after which the work was held for a while because of the floods of August 2018. The floodwater caused significant changes in soil behavior. The surface showed considerable settlement due to the instability of the clayey soil. Also, the water table rose to a shallow depth which was evident at the time of sample collection. The surface of soil was black in color. The top 20–30 cm was rich in organic matter including rotten grasses and roots. For sampling, open pit excavation was adopted at two locations, named MEC1 and MEC2. These locations were at a clear spacing of 15 m. The soil sample was collected after removing 45 cm of topsoil. The pits were of dimension 1.2 m × 0.6 m. Once the excavation was made up to 30 cm depth, water started rising in the pit, indicating the presence of water table at a shallower depth in the area. Approximately 100 kg of soil sample were collected from bottom of each pit in four 50 l plastic containers. All the containers were sealed, labeled (MEC1 and MEC2) and transported to the Geotechnical Engineering (GE) Laboratory at Rajiv Gandhi Institute of Technology (RIT), Kottayam. Also, soil samples were collected for moisture content determination. In situ density of the soil was determined by the core cutter method.

3.2. Basic characterization of the soil

Soil being an important construction material for civil engineers, its testing is essential for any construction project. There have been several cases of failure of projects due to improper or inadequate soil testing. Unlike steel or concrete, whose properties are known for a particular grade, soils exhibit widely varying properties at different places and hence measuring them in the field or in the laboratory is necessary. The tests to be conducted depend on the nature of soil as well as on the type of project.

Tests like moisture content determination, specific gravity test and in situ density test were conducted initially to get an idea about the basic soil properties. The collected soil samples being clayey in nature, Atterberg limit tests and hydrometer tests were used to classify them. The shear parameters of the soil were ascertained using unconfined compressive strength test, direct shear test, and vane shear test. Similarly, the compactness of the cohesive soil was measured in terms of optimum moisture content (OMC) and maximum dry density (MDD) by conducting standard proctor compaction test. The soil seemed black in color and was rich in organic matter and hence organic content and free swell test were performed. Because of the geographical nearness to the Kuttanad region the pH of the soil was also measured. The results of basic characterization tests conducted on soil samples are summarized in Table 1.

Table 1. Basic soil properties.

Properties	MEC 1	MEC 2
Specific gravity	2	2
In situ water content (%)	71.93	85.71
Field density—dry(g/cc)	0.782	0.743
Liquid limit (%)	123	138
Plastic limit (%)	72.97	93.49
Plasticity index (%)	50.03	44.51
UCS (kg/cm^2)	0.054	0.026
MDD (g/cc)	1.304	1.062
OMC (%)	23.08	29.17
Free swell index (%)	10	10
pH	3.4	3.23
Organic carbon (%)	8.08	8.08
Organic matter (%)	13.92	13.92

3.3. Stabilization of the soil

From the basic characterization part, it was understood that the soil possessed very low shear strength, expansive behavior, and acidic nature, which revealed its unsuitability for construction. So, stabilization of the soil was necessary to alter its properties in terms of increase in shear strength and control of the swell shrink tendency. Different methods of stabilization are available, but the soil behavior suggested chemical stabilization to be more suitable. Because of the low cost, easy availability, and faster effect, magnesium chloride was selected. Also, magnesium chloride is environment-friendly and suits the particular clayey soil.

3.3.1. Sample preparation

Oven-dried soil samples were used for studying the effect of magnesium chloride addition so as to have a better control over the water content. The soil samples were oven-dried for 24 h at 105°C, then pulverized and sieved through 2.36 mm sieve, to attain a uniformly mixed soil sample. The soil samples were separated in desired quantities and were added with different percentages of $MgCl_2$ (3%, 6%, 9%, 12%, and 15%) by the weight of soil. The salt in powder form was mixed thoroughly with soil and kept in an air-tight plastic bag for one day curing. The trial mix showed that the salt mixed thoroughly with soil affected its plasticity characteristics. So, the effects were analyzed by conducting liquid limit and plastic limit test. Also, the influence of $MgCl_2$ on the strength of soil was studied by conducting vane shear strength test. For studying the effect of stabilization of soil both untreated (0% $MgCl_2$) and treated soil samples were tested. Three trials were conducted on each soil sample (MEC1 and MEC2) to find the liquid and plastic limits for each percentage addition of $MgCl_2$. The percentage addition at which the least value was obtained or at which the value flattens was considered as optimum percentage of $MgCl_2$. The optimum percentage was considered as that percentage of $MgCl_2$ corresponding to the peak strength.

4. RESULTS AND DISCUSSION

The study of basic soil properties showed a very low shear strength, slight expansive behavior, and the acidic nature. Trials test conducted on soil samples by mixing magnesium chloride in powdered form showed that the stabilizer has some affinity toward the soil as it gets mixed thoroughly. This implies a possible interaction between soil particles and magnesium chloride is happening at micro levels.

4.1. The effect of magnesium chloride on liquid limit, plastic limit, plasticity index, and the shear strength of the soil

The liquid limit of the untreated soil sample was 60% and 67.5% for the sample MEC1 and MEC2. From Figure 1, when 3% of magnesium chloride was added the liquid limit reduced to 54% for both samples. On further increment of magnesium chloride results in continuous reduction in the liquid limit which reaches the value of 45% and 32% at 15% addition of magnesium chloride.

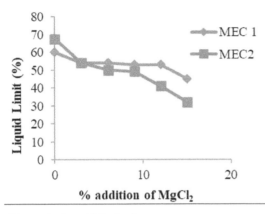

Figure 1. Liquid limit plot.

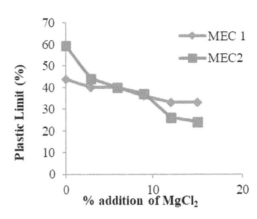

Figure 2. Plastic limit plot.

From Figure 2, the plastic limit of the sample MEC1 and MEC2 was initially 43.71% and 59.4% which reduces to 40% and 44% with the addition of 3% magnesium chloride. The plastic limit reduces further by the addition of increased amount of magnesium chloride.

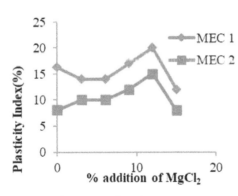

Figure 3. Plasticity index plot.

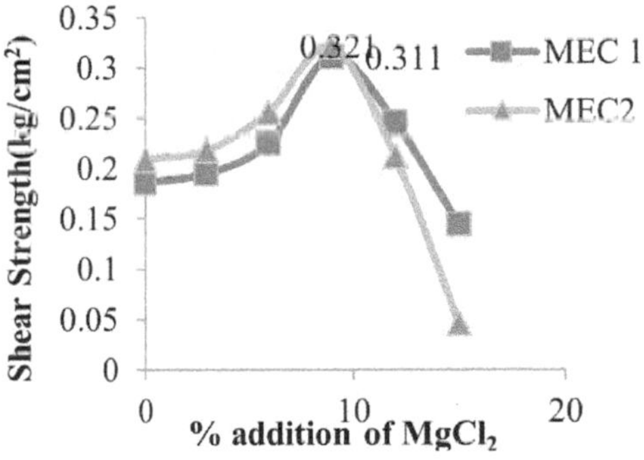

Figure 4. Vane shear strength plot.

From Figure 3, the plasticity index of the soil for the addition of magnesium chloride is found using the liquid limit and plastic limit. The peak of plasticity index value for different percentage addition is obtained at 12% which can be considered as the optimum percentage of addition of magnesium chloride for MEC1 and MEC2. The decrease in liquid limit and plastic limit is due to the reduction in the thickness of diffuse double layer as the magnesium chloride content is increased. The introduction of magnesium ion in to the soil will replace the intermediate and outer water layers of diffused double layer. The property of the pore water changes which results in the change of plasticity characteristics (Murthy, 1995).

The soil samples were tested by laboratory vane shear test for one day curing and at water content of OMC+20% for MEC1 and OMC+15% for MEC2. This water content was used for all the percentage addition of magnesium chloride. The initial strength of untreated soil samples MEC1 and MEC2 was found to be 0.186 and 0.209 kg/cm² which gradually increased with the addition of magnesium chloride up to 9% as shown in Figure 4. Further increase of the magnesium chloride content reduced the shear strength of the soil in both samples. Both the samples obtained a peak at 9% addition of magnesium chloride by weight of soil which can be considered as the optimum percentage to obtain an increase in strength of the soil. Magnesium chloride is having base-exchange capacity which influence the plasticity characteristics of the soil. The changes in the plasticity characteristics result in the strength character of the soil when magnesium chloride is used. The higher percentage of magnesium chloride addition will result in release of more water from the adsorbed layer which hinders the lubrication and hence the interaction of soil particles. This will reduce the strength of soil when excess amount of magnesium chloride is used (Ayyar, 2000). The usage of magnesium chloride in the case of fine soil with high organic content with low value of pH improves the strength behavior of soil. The increase in the strength of the samples agrees on the use of magnesium chloride as an effective stabilizer.

4.2. Optimum percentage

The optimum percentage of $MgCl_2$ to be added shall be taken as the average of values obtained from the plasticity characteristics and strength characteristics. The peak value of plasticity index was obtained at 12% addition of magnesium chloride for both samples. The peak value of the strength characteristics was obtained at 9% addition of magnesium chloride in case of both samples. The optimum percentage of magnesium chloride can be taken as 10% addition of magnesium chloride. The experimental results suggest that the soil with the addition of 10% $MgCl_2$ by weight is expected to show an optimum behavior of strength improvement and plasticity characteristics of organic soils of low pH value with slight expansive behavior.

5. CONCLUSIONS

The untreated soil had low strength, high organic content, slightly expansive, acidic nature, and very high liquid limit and plastic limit. For stabilization of the soil, magnesium chloride was selected. The optimum percentage of additive ($MgCl_2$) for plasticity character improvement is 12% as the plasticity index has maximum value. The soil sample MEC1 with OH classification that is organic soil with high compressibility has shifted to the left and has crossed the line of 50% liquid limit to become soil of intermediate compressibility (OI). The soil sample MEC2 shifted to OL (organic soil of low compressibility) from OH. Optimum percentage of additive for strength gain is 9% as it shows maximum shear strength in the case of both samples. The soil with the addition of 10% magnesium chloride by weight is therefore expected to show an optimum behavior of strength improvement and plasticity characteristics.

REFERENCES

[1] Arora, K. R., *Soil Mechanics and Foundation Engineering*, seventh edition, Standard Publishers, 2015.

[2] Gregory Paul Makusa, "Soil stabilization methods and materials," Department of Civil, Environmental and Natural resources Engineering, State of the Art Review, 2012.

[3] Juhaizad Bin Ahmad, "Improving the stability of clay for construction," 2014. www.sciencedaily.com.

[4] Karl Terzaghi, Ralph B. Peck, and Gholamreza Mesri, *Soil Mechanics in Engineering Practice*, third edition, Wiley-India, 2010.

[5] Murthy, V. N. S., *Soil Mechanics and Foundation Engineering*, fourth edition, SAI Kripa Technical Consultancy, 1995.

[6] Nazile Ural, "The importance of clay in geotechnical engineering," 2018. www.intechopen.com.

[7] Onyelowe and Kennedy Chibuzor, "Soil stabilisation techniques and procedures in the developing countries – Nigeria," *Glob J Eng Technol*, vol. 5, pp. 65–69, 2012.

[8] Ramanatha Ayyar, T. S., *Soil Engineering in Relation to Environment*, first edition, LBS Centre for Science and Technology, 2000.

22. Stabilization of Clayey Soil by the Blend of Rice Husk Ash and Chicken Bone Powder

Ketan Chaudhary* and Preetpal Singh

Department of Civil Engineering, Rayat-Bahra University, Mohali, Punjab, India

*Corresponding author: ketanchaudhary87@gmail.com

ABSTRACT: The quality of soil and other resources is declining due to pollution, which has an impact on building due to unfavorable findings from testing programs. Soil beneath a structure is necessary for its stability. Therefore, improving soil quality is necessary in order to develop sturdy structures and buildings. There are already various ways to improve soil properties, but they need a lot of time and work. They are not therefore affordable or practical. The current inquiry uses two products, rice husk ash (RHA) and chicken bone powder (CBP), with the aim of improving soil qualities. These two products, which are generated in large quantities annually by two separate industries, not only strengthen soil when mixed with it, but also address the issue of how to dispose of them. RHA is a rice huller derivative that exhibits pozzolanic capabilities and contains an 83% silicon dioxide concentration in its main component. However, CBP, which contains roughly 55% of calcium phosphate, has a strong heat transfer resistance. Therefore, clayey soil to be stabilized is gathered from village Dariya Ke, Punjab, and various experiments were performed on clayey soil with the purpose of enhancing the performance and features of clayey soil by use of waste products such as RHA and CBP. The composite soil sample is used to improve the characteristics of clayey soil and simultaneously an initiative to control pollution by implementing waste products in construction techniques. The composite soil sample is created by blending the substances in varying quantities of RHA as 1%, 3%, 5%, and 7% and as that of CBP about 2%, 4%, 6%, and 8%. To assess the strength behavior of soil UCS and California bearing ratio (CBR) test are carried out. The results obtained by mixing the materials in different proportions showed that the optimum proportion of RHA as 5% and CBP as 6% assists in enhancing the strength parameters. The UCS increased from 79.0 kPa (virgin soil) to 289.7 kPa and CBR shows a rise in experimental value from 3.25 to 7.28. Thus the results reveal that with the usage of composite soil, the characteristics of soil like UCS can be enhanced and thus making it fit for construction purposes.

KEYWORDS: Clayey Soil, Rice Husk Ash, Chicken Bone Powder, California Bearing Ratio, Unconfined Compressive Strength

1.   INTRODUCTION

The characteristics of soil start fluctuating as we proceed from one location to other. The stability of a structure or building depends upon the soil underneath it. So for a structure to have a strong foundation and to be firm righteous kind of soil is needed, but it is not feasible to get the required soil which fulfills the necessary parameters for construction purposes at every place. To tackle this issue of unavailability of apt soil at some places we modify the characteristics of soil to stabilize it and make it felicitous for use. But these modifications in the quality of soil increases the overall cost of project and thus making it uneconomical. Moreover, it becomes a challenge for civil engineer to accomplish the project as per given guidelines and in specified amount of time depending upon type of construction. The crucial

Chapter 22 DOI- 10.1201/9781032657271-22

aspect is detection of clayey soil in the most of the regions which possess high water retention and poor permeability which causes potential swelling or potential shrinkage of due to the companionship of minerals namely kaolinite, illite, and montmorillonite. So stabilization of soil occupies a considerable place in civil engineering, as we can resolve many issues faced during construction by the use of various materials available especially byproducts from industries, agriculture, etc. So this approach not only helps in enhancing soil quality, but also in waste management, therefore reducing pollution. So the present study deals with the implementation of rice husk ash (RHA) and chicken bone powder (CBP) with the aim to ameliorate the soil characteristics and thus increasing strength.

Rice hulls are the exterior rigid covering of the cereal of rice which preserves the rice during its augmentation time and are not ingested by human beings thus, due to non-consumption it becomes a farming residue left in prodigious quantity every year about 520–670 Mt annually thus, becoming a crucial aspect of waste regulation contaminating the atmosphere due to scanty obtainability of area for disposing purposes. RHA, which is extracted during the combustion of rice hulls at a high firing point of about 773.15–973.15 K, contains an abundant amount of silica dioxide of about 82–85%. Silica dioxide possesses pozzolanic properties, which makes it fit for supplementary binding material. So RHA in different proportions of 1%, 3%, 5%, and 7%, respectively. The impervious microstructure of RHA claims greater resistance to sulfate attack, carbonation, etc. As RHA is very lightweight it will not cause any difficulties in compaction. It is also observed from past research that it has a very high angle of internal friction depicting very high stability and also manifest good shrinkage properties.

Chicken bone powder which is a waste of poultry left after the use of skin of birds containing calcium phosphate (53–55%) and phosphorus pentoxide (40–42%) and magnesium oxide, carbon dioxide, silicon dioxide, iron(II) oxide, aluminum(II) oxide is mixed with virgin soil in 2%, 4%, 6%, and 8%, respectively. CBP has high resistance of heat transfer and it increases shear strength, California bearing ratio (CBR) of soil. When RHA (which majorly contains silica dioxide) and CBP (which majorly contains calcium) are mixed, a chemical reaction happens between them and they bind together which results in very fine stabilization.

With reference to past studies of various researchers, we came to know that by the use of rice husk ash at the optimum value liquid limit, plasticity index, delineate a downswing (Fattah et al., 2013). Suneel Kumar and Preethi (2014) in their study revealed that by inclusion of the mixture in the soil the CBR and UCS values surged. Utami (2014) used lime and their studies depict that CBR portrays an increment up to an optimum value and LL and PI values decreased. Karthik et al. (2014) made use of fly ash and CBR values showed an increment, which consequently resulted in an increment of UCS at the most desired content of material. Anu Paul (2014) used ESP and quarry dust and deduced that by the substitution of these substances the soil properties were enhanced. Singhai et al. (2014) avails RHA and Lime and outlined that differential swell showed a decrement. Sruthimol (2015) utilized Baggase ash and portrayed an enrichment in characteristics of soil. Ajay Upadyhya (2016) used ceramic waste and a decrement in LL, PL, and PI. Otoko et al. (2016) published his study in which they portrayed the use of Palm oil fiber in soil and there was an enhancement in soil properties. Onyelowe (2016) made the use of Bone powder in soil and MDD values were increased and it also enhanced the cohesion. Alrubaye et al. (2016) made use of lime and his study revealed that Kaolin, a typical type of clay, got stabilized. Butt et al. (2016) used saw dust and the CBR showed an increment and consequently, UCS was also increased. Athira (2017) mixed coconut shell powder and lime in soil and an increase in compressive strength was noted. Wilson and Sudha (2017) showed that properties of clayey soil get enhanced by make use of glass powder and groundnut shell ash. Jagtar and Rattan (2017) by mixing shredded rubber tire concluded that a rise in OMC was seen due to the water-retaining feature of rubber and also the usage of shredded rubber tire is also helps in reducing pollution. Goyal and Sharma (2018) used

Recron-3S and fly ash and infer that geo fiber of the new era which not only increased the OMC but also an increment in UCS was seen.

Ramesh and Hima Bindhu (2019) used blast furnace slag and a gentle increment in OMC along with MDD was observed. Adeyanju (2019) made use of cement kiln dust and an increment in workability was noticed along with decrement in project cost as the depth of pavement is lessened. Hidalgo (2019) used SBA and RHA and at the optimum proportion of mixture MDD, OMC portrayed an increment. Hassan and Sharma (2019) in his study mixed plastic strips waste and cement kiln dust which not only stabilized soil but also the usage of plastic stripes helps in reducing pollution and thus this method is eco-friendly too. Abhishek (2020) made use of construction and demolition waste and study revealed that drainage characteristics of the soil were enhanced but on the other hand MDD portrayed a diminished pattern. Singh (2021) used a mixture of POP and groundnut shell ash and properties of soil got enhanced along with an increase in shear strength. Kumar et al. (2021) observed a drop in swelling pressure and swelling potential when a mixture of geocell and jute fiber was used.

2. MATERIALS USED

2.1. Clayey soil characterization

For the accomplishment of the research the soil is acquired from the field located in Village Dariya Ke in Ferozepur region of Punjab, India and several testing procedures were conducted namely liquid limit, plastic limit, plasticity index, specific gravity, SPT, UCS, and CBR to determine the basic physical properties of virgin sample of soil and the results obtained are illustrated below in tabular form as:

Table 1. Physical properties of virgin soil.

Tests performed	Results	
Wet sieve analysis	57% virgin soil is passing from 75 micron sieve, so it is a fine grain clayey soil	
Liquid limit (LL)	41.7%	
Plastic limit (PL)	23.4%	
Plasticity index (PI)	18.3%	
Specific gravity	2.38	
Standard proctor test (SPT)	OMC	MDD
	14.96%	1734 kg/m^3
Unconfined compressive strength (UCS) (kPa)	79	
California bearing ratio (CBR) (%)	3.25	

2.2. Rice husk ash

Rice hulls which cover the outer periphery of rice crop, mainly consisting of silica, lignin, and opaline are used as burning fuel in boilers, power plants, various industries produce a large amount of ash which is very particulate in nature and proper disposal of RHA requires enormous amount of land and effort as well. Thus improper management and disposal causes pollution to a great extent. But using it as a stabilizing material not only reduces the waste disposal, but also makes the project economical without any compromise in required parameters, as it holds immense amount of silica possessing pozzolanic properties that helps in stabilization of soil. But it cannot be used alone as it lacks calcium constituent. RHA to be utilized is acquired from Friends Rice Mill, Gidderbaha, Punjab, and the data of chemical composition of RHA is gathered from Suneel Kumar and Preethi (2014), respectively. The tabular form shown below portrays the physical properties, chemical composition of RHA:

Table 2. Physical properties of RHA.

Particulars	Properties
Color	Blackish brown
Shape texture	Irregular
Particle size	Less than 45 micron
Odor	Odorless
Specific gravity	2.3
Appearance	Very fine

Table 3. Chemical composition of RHA.

Constituents	Percentage (%)
Silicon dioxide	83.60
Aluminum oxide	3.5
Ferric oxide	1.10
Calcium oxide	1.80
Magnesium oxide	1.28
Sodium oxide	0.17
Potassium oxide	0.29

2.3. Chicken bone powder

The bones of birds used are collected from meat shop which is situated in Gidderbaha, Punjab that are left behind after the use of skin comprising mainly of calcium phosphate and phosphorus pentoxide helps to stabilize poor soil. Bone powder to be utilized for study is gathered by performing the following steps:

1. Firstly, the accumulated bones are kept in open atmosphere so, that the moisture evaporates out.
2. In the second step, the bones are burnt so that the hygroscopic moisture evaporates out and are left down to cool.
3. In the third step, bones are pulverized in a mixer to get the required ash and kept in fully protected area away from moisture, rain, and heat. CBP chemical composition is acquired from G.M. Journal of Emerging Trends in Engineering and Applied Science and noted down in Table 4.

Table 4. Chemical composition of chicken bone powder.

Constituents	Percentage (%)
CaO	55.60
P_2O_5	41.65
MgO	1.40
CO_2	0.43
SiO_2	0.08
FeO	0.07
AlO	0.05

3. METHODOLOGY

With perspective to enhance the characteristics and ameliorate the properties of clayey soil by the employ of the blend of RHA and CBP various testing programs were conducted and geophysical characteristics were found out. The various processes involved were wet sieve analysis as per IS:2720(Part-4):1985, Atterberg's limit test IS:70(Part-5):1985, and specific gravity test IS:2720(Part-40):1977. These all testing procedures were initially conducted on the virgin soil and after that firstly on RHA with different combination and secondly on CBP and finally on the composite sample as per IS norms. To find out the compaction characteristics as per IS: 2720(Part-7):1980, unconfined compressive strength IS: 2720(Part-10):1991 and CBR IS: 2720(Part-16):1987 a number of soil samples were collected with different proportion of materials to be added as shown in Table 5.

Table 5. Mixed proportions of the composite soil sample.

Description	% age of particulars
CS:CBP:RHA	98:2:0
CS:CBP:RHA	96:4:0
CS:CBP:RHA	94:6:0
CS:CBP:RHA	92:8:0
CS:CBP:RHA	99:0:1
CS:CBP:RHA	97:0:3
CS:CBP:RHA	95:0:5
CS:CBP:RHA	93:0:7
CS:CBP:RHA	97:2:1
CS:CBP:RHA	93:4:3
CS:CBP:RHA	89:6:5
CS:CBP:RHA	85:8:7

4. RESULTS AND DISCUSSION

Table 6 portrays the results derived by substitution of different percentage of RHA and CBP to discern various parameters including optimum moisture content (OMC), maximum dry density (MDD), and unconfined compressive strength (UCS) as follows.

Table 6. OMC, MDD, and UCS results at varying proportion of rice husk ash and chicken bone powder.

RHA (%)	CBP (%)	OMC (%)	MDD (kg/m³)	UCS (kPa)
0	0	14.96	1734	79.0
0	2	14.16	1785	96.7
0	4	14.01	1826	138.3
0	6	13.76	1903	207.4
0	8	13.18	1846	189.5
1	0	15.15	1751	93.5
3	0	16.24	1789	127.0
5	0	16.36	1823	223.5
7	0	15.98	1787	209.8
1	2	14.56	1806	238.4
3	4	15.03	1829	254.6
5	6	**15.36**	**1877**	**289.7**
7	8	15.47	1853	277.3

It can be clearly understood from the tabular form represented in Table 6 that by the inclusion of RHA and CBP the strength parameters show an increment in results. Addition of RHA to clayey soil resulted in increment of stability of soil due to high angle of internal friction of RHA. On the other side, substitution of CBP reduces voids in soil due to high specific gravity of bones and also a fall in OMC due to permeability property of CBP thus enhancing strength parameters.

Table 7. CBR observations at different percentage of CBP and RHA.

Proportion of CBP (%) and RHA (%)		CBR (%)
	2.5 mm	5 mm
Virgin clayey soil	3.25	3.17
2% CBP	4.09	3.84
4% CBP	5.69	5.18
6% CBP	6.78	6.29
8% CBP	6.14	6.02
1% RHA	4.19	3.98
3% RHA	5.24	4.63
5% RHA	6.91	6.57
7% RHA	6.17	5.94
1% RHA + 2% CBP	5.79	5.16
3% RHA + 4% CBP	6.94	6.04
5% RHA + 6% CBP	**7.28**	**7.05**
7% RHA + 8% CBP	7.08	6.97

From Table 7, it can be observed that the penetration results of CBR obtained on the substitution of CBP and RHA in different proportions portray unique results of CBR, giving us quite satisfactory values and depicting the highest value of CBR at the optimum range of 5% RHA + 6% CBP, thus an amplification in soil properties were observed.

5. CONCLUSION

The following results can be concluded from the values obtained by the use of varying quantities of RHA and CBP as following:

1. The combination of 5% of RHA and 6% of CBP came out to be an optimum value which yields quite satisfactory results by manifesting a growth in strength variables of the clayey soil, any further augmentation in the content of the substances used will illustrate a decrease in the values calculated by the use UCS and CBR testing machines.
2. An enormous increase in UCS is observed as its value rises from 79 to 289.7 kPa which is approximately 3.6 times the value of virgin soil and the report obtained from CBR testing machine also gives quite interesting results with a huge rise in CBR values.
3. Rice hulls that are generated on large magnitude almost everywhere on earth becomes a reason of strain due to its substantial role in causing pollution, by the execution of RHA in unwanted clayey soil with aim to enrich the clayey soil characteristics we are resolving two issues firstly an positive step to refine the soil properties and secondly an initiative to dispose off the RHA, assisting in management of waste and thus we are hitting two targets with a single arrow.

REFERENCES

[1] Abhishek, S., *Strength and Drainage Characteristics of Poor Soils Stabilized with Construction Demolition Waste*, National Institute of Technology, Hamirpur, Himachal Pradesh, India, 2020. https://doi.org/10.1007/s10706-020-01324-3

[2] Adeyanju, E. A., "Clay soil stabilization using cement kiln dust," *1st International Conference on Sustainable Infrastructural Development IOP Conf. Series: Materials Science and Engineering 012080*, IOP Publishing, 2019, doi: 10.1088/1757-899X/640/1/012080.

[3] Ajay, U., "Review on soil stabilization using ceramic waste" *Int Res J Eng Technol*, vol. 03, no. 07, pp. 1748–1750, 2016.

[4] Alrubaye, J. A., et al., "Engineering properties of clayey soil stabilized with lime," *ARPN J Eng Appl Sci*, vol. 11, no. 4, 2016.

[5] Amiralian, S., et al., "A review on the lime and fly ash application in soil stabilization," *Int J Biol Ecol Environ Sci*, vol. 9, pp. 124–126, 2012, doi: 10.1007/s10706-020-01366-7.

[6] Anil, P., et al., "Studies on improvement of clayey soil using egg shell powder and quarry dust," *Int J Eng Res Appl*, vol. 4, no. 4, pp. 55–63, 2014, doi: 10.1016/j.jafrearsci.2019.01.

[7] Butt, W. A., Gupta, K., Jha, J. A., "Strength behavior of clayey soil stabilized with saw dust ash," *Geo-Engineering*, vol. 7, Article no. 18, 2016. doi: 10.1186/s40703-016-0032.

[8] Fattah, M. Y., Rahil, F. H., Al-Soudany, K. Y. H., "Improvement of clayey soil characteristics using rice husk ash," *J Civil Eng Urban*, vol. 3, no. 1, pp. 12–18, 2013.

[9] Goyal, T., Sharma, Er. R., "Experimental study of clayey soil stablised with fly ash and RECRON-3S," *Int Res J Eng Technol*, vol. 05, no. 10, pp. 724–729, 2018. doi: 10.1080/10962247.2020.1862939.

[10] Hassan, S., Sharma, N., "Strength improvement of clayey soil with waste plastic strips and cement kiln dust," *Int Res J Eng Technol*, vol. 06, no. 11, 2019.

[11] Hidalgo, F., "Stabilization of clayey soil for subgrade using rice husk ash (RHA) and sugarcane bagasse ash (SCBA)," *IOP Conf. Series: Materials Science and Engineering*, 758012041, ICMEMSCE, doi:10.1088/1757-899X/758/1/012041.

[12] Jagtar, S., Rattan, J. S., "Soil stabilization of clayey soil using shredded rubber tyre," *Int J Eng Res Technol*, vol. 6, no. 9, paper ID IJERTV6IS090113, 2017.

[13] Karthik, S., Kumar, A., Gowtham, P., Elango, G., Gokul, D., Thangaraj, S., "Soil stabilization by using fly ash", *IOSR J Civil Mech Eng*, vol. 10, pp. 20–26, 2014.

[14] Kumar, S., Sahu, A. K., Naval, S., "Study on the swelling behavior of clayey soil blended with geocell and jute fibregeocell and jute fibre," *Civil Eng J*, vol. 7, no. 8, pp. 1327–1340, 2021. doi: 10.28991/cej-2021-03091728.

[15] Onyelowe, K. C., "Kaolin stabilization of olokoro lateritic soil using bone ash as admixture," *Int J Constr Res Civil Eng*, vol. 2, no. 1, pp. 1–9, 2016.

[16] Otoko, G. R., Fubara-Manuel, I., Chinweike, I. S., Oluwadare, J., Oyebode, O., "Soft soil stabilization using palm oil fibre ash," *J Multidiscip Eng Sci Technol*, vol. 3, no. 5, pp. 2458–9403, 2016.

[17] Ramesh, B., Hima Bindhu, P., "An experimental study on stabilization of clayey soil by using granulated blast furnace slag," *Int J Trends Sci Res Dev*, vol. 3, no. 5, pp. 655–658, 2019.

[18] Satnam, S., Singh, G., "Stabilization of clayey soil using waste plaster of Paris and groundnut shell ash," *Int J Eng Res Appl*, vol. 11, no. 1, pp. 17–24, 2021. doi: 10.9790/9622-1101041724.

[19] Singhai, A. K., "Stabilization of soil with rice husk ash and fly ash," *Int J Res Eng Technol*, vol. 5, pp. 545–549, 2014.

[20] Sruthimol, P., "Experimental study on the effect of bagasse ash," *Int J Eng Res Technol*, vol. 4, no. 11, 2015.

[21] Utami, G. S., "Clay soil stabilization with lime effect the value CBR and swelling," *ARPN J Eng Appl Sci*, vol. 9, no. 10, pp. 1744–1748, 2014.

[22] Suneel Kumar, B., Preethi, T. V., "Behavior of clayey soil stabilized with rice husk ash & lime," *Int J Eng Trends Technol*, vol. 11, no. 1, pp. 44–48, 2014.

[23] Wilson, E. S., Sudha, A. R., "A comparative study on the effect of glass powder and groundnut shell ash on clayey soil," *Int J Eng Res Technol*, vol. 6, issue 2, 2017.

23. Remediation Techniques on Red Mud—A Review

Jinsha T V [1,*] and Beena K S [2]

[1]Division of Civil Engineering, Cochin University of Science and Technology, India

[2]SOE, Cochin University of Science and Technology, India

[*]Corresponding author: jinsha.t.v@gmail.com

ABSTRACT: As land is scarce due to population explosion, safe disposal of industrial by-products (IBPs) is one of the significant challenges we face nowadays. It will be a great relief if IBPs can be decontaminated after heavy metals extraction or lowering pH and use effectively as a geomaterial. So, there is a rising demand for new technologies to accelerate the waste products decontamination and valorize the IBPs at a reduced cost to meet conventional resources' insufficiency for construction purposes. Red mud (RM) is a bauxite residue formed during alumina's production by Bayer's process when the bauxite ore is subjected to caustic leaching. As India is among the significant alumina producers globally and more than 4 million tons of RM is being generated annually, high long-term storage of RM remains the economic and environmental management challenges of aluminum industries. Its applicability as secondary raw material is hindered mainly by its extreme pH value, potentially toxic elements and radionuclides present, affecting its multiple engineering properties. So, if the RM can be ameliorated by innovative sustainable remediation technologies that would follow the principles of the circular economy in the mining sector, RM disposal issues can be mitigated to a large extent. This study critically reviews various remediation technologies of RM of different origins and their beneficial use as a sustainable material.

KEYWORDS: RM, Geomaterial, Decontamination, Remediation

1. INTRODUCTION

Rapid industrialization and faster growth lead to a sustained increase in the living standard of people, but for sustainable development, environmentally sound industrialization is essential. One such example is the issue of the bauxite residue or red mud (RM) from the global aluminum sector. In the manufacture of aluminum from bauxite, the unsolvable components of the bauxite are removed by digesting the ore with very hot caustic soda (sodium hydroxide) and the unreacted slag in bauxite, which is the waste material in the Bayer process, is called "red mud." RM includes radioactive and hazardous heavy metals. Due to its high alkalinity, it is exceedingly caustic and harmful to soil and biological forms, creating a significant disposal issue (Mayes et al., 2016; Mukiza et al., 2019). With annual world aluminum production expected to top 117 million tons per year, the continuously increasing production of every year, RM produced as a by-product poses a serious hazard to the environment (Putrevu et al., 2021). Technology currently provides the means to store the residue in a controlled manner; however, this does not completely address the issue. Every bauxite/alumina facility in the world has toxic settling pools and dumps near to it. The leading aluminum companies are investigating various strategies to improve the situation in the wake of major disasters, including the Bauxite residue accident from Ajka, Hungary, the Norwegian Hydro in Brazil in 2018, and the RM pond dam failure in China's Henan Province in 2016. Aging containment ponds and anticipated demand growth for aluminum have urged the industries to look for better solutions for the effective valorization of the RM (RM Project, 2021). While RM composition varies based on the source of bauxite and other variables, it is comprised of various oxides, mainly iron oxide content (which gives the RM its name). The main barriers that prevent RM from being used in bulk are its complicated mineralogical and chemical composition and other physical–chemical characteristics such as excessive fineness, poor settling, and dewatering characteristics. Because of the significant disparity between the production and consumption of RM at the current technological stage and practice, full utilization of RM remains a concern on a worldwide scale (Rai et al., 2020). A single technology for the remediation of the contaminants may not be that appropriate for selecting RM as a suitable geomaterial. Thus, it is crucial to have an in-depth understanding of several technologies to combine them to overcome problems associated with using a particular method.

2. METHODOLOGY

The use of RM in various fields was previously discussed in several review studies. There is a substantial divide between the production and use of RM at the present state of technology and practice. The entire usage of the RM is still an issue on a global scale. Hence, the current study focuses on various remediation technologies applied to the RM for its bulk utilization. Recycling industrial waste into building and construction materials by mixing it with adhesives (such as cement or lime) exposes inhabitants, labors, and other people who belong to the general population to danger because of the increased release of gamma rays from the recycled materials (Sahoo and Joseph, 2021). Therefore, to tackle the bulk quantity of waste generation, environmentally benign, economically feasible, the time and energy-intensive methodologies for remediation without the generation of secondary pollutants need to be found out for its wide acceptance among most techniques. Here, literature featuring many experiments on RM globally is evaluated critically. Researchers will better understand the various remediation technologies as a result of the review presented here and will be able to utilize them to solve RM-related issues in various applications.

3. REMEDIATION METHODS

3.1. Stabilization

Stabilization is a remediation technology that involves the addition of reagents or inorganic cementitious binders to the waste to transform it in to a new solid material which is more chemically stable. Stabilization of RM by varying percentage of lime to improve its geotechnical properties for its effective utilization as backfill subgrade material and clay liners has been explored by several researchers. Kalkan (2006) examined the feasibility of using cement–RM additives as hydraulic barrier in compacted clay liners. Based on the test results, compacted clay with RM and RM with cement additives had more compressive strength, reduced hydraulic conductivity, and a lower percentage of swelling than samples of natural clay. The addition of these chemicals altered the soil groups from the high-plasticity soil group (CH) to the low-plasticity soil group (MH) (Kalkan, 2006). Satayanarayana et al. added lime to RM at varying percentages (2–12%) and observed the changes in its geotechnical properties(unconfined compressive strength [UCS], split tensile strength, and California bearing ratio) after curing for 1, 3, and 28 days. It was found that 10% of lime had outperformed other percentages. And CBR value was 25% after 28 days (Satayanarayana et al., 2012). Subsequently, Deelwal et al. (2014) added gypsum (1%), along with lime (4%, 8%, and 12%), and found that the combination of 12% lime with 1% gypsum had a more significant UCS value after curing for 7 days and higher value of CBR, that is, 7.9% in comparison to other percentages. Moreover, it was discovered that permeability decreased when lime and gypsum concentration increased (Deelwal et al., 2014). As a result, RM stabilized with lime, either with or without gypsum, can be used as a filler in the building of embankments, subbases, base courses, and subgrades for roads. According to Kushwaha et al., when RM is stabilized using Eko-Soil (ES) enzyme, the addition of ES (up to 4%) improves maximum dry density and plasticity but decreases optimum moisture content, liquid limit, and permeability. When gypsum and lime are added to the

RM–ES mix, both the soaked and unsoaked CBR values increase significantly (Kushwaha et al., 2018). Also, when compared to other combinations, the 4% ES heavy metal content is relatively low. It can be utilized for many geotechnical applications, including building embankments, roads, dams, reservoirs, and mine filling.

3.2. Bioremediation

Many studies on the effectiveness of RM remediation utilizing microbes, fungi, biopolymers, etc. have already been conducted in order to lessen the adverse effects of RM, such as its high alkalinity, the presence of hazardous components, and its leachability when used as a geomaterial. Aspergillus Niger was used in a study by Qu et al. (2013) to examine the bioleaching of heavy metals from RM. The bioleaching tests were conducted in batch cultures with the RM at different pulp densities ranging from 1% to 5% (w/v), and it was shown that spent medium leaching at 1% pulp density produced the highest leaching ratios of the majority of different heavy metals. With the increase in RM pulp densities, there has been a general decrease in leaching ratios under all bioleaching conditions. Using both indigenous and non-indigenous bacteria like Lactobacillus acidophilus, NCIM 2660, and *Lactobacillus plantarum* NCIM 2083, Panda et al. (2017) assessed the effects of bio-neutralization on the morphological, physical, and geotechnical features of RM (LAB 4). It was found that the RM's consistency has improved in terms of plasticity index and compaction characteristics due to its low optimal moisture content. When compared to untreated RM, the UCS value increased to 298.6 from 136.5 kPa. The data analysis shows that there are several prospects for employing RM in considerable amounts as a fill material and an embankment material after bio-treatment. Lavanya and Kumar (2017) conducted studies on the RM with two different biopolymers. Xanthan gum (XG; 0.5%, 1%, and 2%), and guar gum (GG; 0.5%, 1%, and 2%) and found that the addition of XG and guar gum biopolymer to RM increases OMC values and decreases MDD values. Reddy et al. (2018) investigated ways to reduce the dispersion characteristics of RM waste using eco-friendly biopolymers like GG and XG, in dosages of 0–5, 1%, 2%, 3%, and 4% (by weight). They discovered that as the dosage of the biopolymer has shown increase in both the liquid limit and PI, while the plasticity limit remained nearly constant (Reddy et al., 2018). Also, they noted that GG is a superior stabilizer for reducing the dispersion and dusting behavior of RM Waste while XG is ineffective. To develop an environmentally acceptable way to solidify RM, the use of MICP technology has recently been promoted, and it has been found that this technology is highly exploited in RM and is capable of neutralizing excessive alkalinity (Liu et al., 2021). They also treated Bayer process RM through MICP and it was found that the RM got much stronger, with a change from strain hardening to strain softening with an increase of UCS up to 1395 kPa. As part of the treatment, the pH could be lowered, and the heavy metal ions in the RM could be precipitated as carbonate. Mechanistic analysis says that the only thing that changed the RM was the calcite, which had a huge volume and was getting aggregated. Cementation caused the RM particles to combine and transform from a loosely dispersed state to a compact and solid state, resulting in an increase in strength and UCS. Additionally, the treatment could potentially lead to the immobilization of heavy metal ions, making the RM safer for disposal or reuse.

3.3. Phytoremediation

It is a difficult task to remediate RM by phytoremediation due to excessive alkalinity, sodicity, elevated amounts of toxic metals, low water retention and nutrient supplying capacities. Prior to the growth of vegetation on RM dumps, strategies for revegetation must be developed. These include field neutralization, soil capping, and the use of organic and inorganic amendments like animal manures, sewage sludge, mushroom compost, vermicompost, and sawdust. Due to salinity and high levels of toxic metals, these amendments provide a favorable substratum for revegetation, although not all plant species may necessarily flourish there. Thus, the potential for phytoremediation of RM deposits through various types of amendments was assessed for a number of plant species which are halophytes, can resist drought, and tolerate metal to a great extent. It was discovered that the growth of the plant *Arundo donax* L. (giant reed) reduces the amount of accessible Cd, Pb, Co, Ni, and Fe. This plant was utilized in studies on the phytostabilization of RM to absorb trace metals and lower pH and salinity. Besides that, the biomass of giant reed seedlings in RM increased by 40.4% and the mud/control soil mixture by 47.2% compared to control soil. The plant's capacity to accumulate metals in the stalk and leaves above the root concentration points to its potential for use as a phytoremediation. Giant reed has improved a number of soil quality indicators, including pH, EC, OC, microbial counts, and soil enzyme activity (Alshaal et al., 2013). A study employing lemongrass grown in soil containing varied amounts of cow dung or sewage sludge (biowastes) revealed that 5% RM combined with biowastes, preferably SS, can be employed as a soil quality enhancer that lowers metal toxicity. Lemongrass works as a potential phytostabilizer of Fe, Mn, and Cu in roots as well as an effective accumulator of Al, Zn, Cd, Pb, Cr, As, and Ni, according to translocation and bioconcentration factors (Gautam and Agarwal, 2017). The phytoremediation of contaminated sites seems to be especially successful with hyperaccumulator plants. In addition to assisting with remediation of contaminated sites, vegetation grown on RM deposits also benefits society economically by improving soil quality, storing carbon, protecting biodiversity, enhancing the esthetics of the landscape, and producing food and medicine (Mishra and Pandey, 2018).

3.4. Neutralization method

Numerous admixtures, including carbon dioxide, seawater, and other acids, have already been used by a number of refineries to try and improve the RM waste (Nikraz et al., 2007; Rai et al., 2012). For the purpose of conducting batch investigations on fluoride adsorption from an aqueous solution, seawater-neutralized bauxite (SWABR) was activated. Fluoride absorption from water increased as the SWABR dosage was enhanced. This was attributed to the SWABR nanoparticles (iron and aluminum) in the bauxite residue having more surface area following activation, which allowed for greater adsorption from ground water samples. Between pH 4 and pH 5, the greatest elimination of fluoride was seen (Chate et al., 2018). The ability of organic acids like oxalic acid and citric acid, as well as the impact of lime on the neutralization of RM, were investigated by Mishra et al. in 2020, taking into account the economy of the technique as well as the transportation and handling of additives because oxalic acid and citric acid require relatively less monitoring and skill for use at the site (Mishra et al., 2020). Furthermore, it was found that no hazardous by-products are produced when organic acids react with RM, making it safe to employ the treated RM in a variety of geotechnical and agricultural applications.

3.5. Electrokinetic remediation

Several studies were also done to determine whether electrokinetic treatment was effective for RM reclamation. To determine the major variables and methods influencing the removal of ions from RM, a number of laboratory electrokinetic investigations are necessary (Sipayung et al., 2017). After considering electrokinetic as a viable remediation method, it was discovered that a zone of pH (8.6) was discovered close to the anode after 30 days. Among the various cations, Na+ removal efficacy (98.01%) was reached in this area with 45 V current. Redmud with a lesser pH can be used as a sustainable geomaterial.

3.6. Thermal treatment

During the heat treatment of RM, the phase transition occurs, which can be studied using XRD and TG–DT analysis. Wu and Liu (2012) studied thermal treatment on RM and the results showed that gibbsite decomposes into Al_2O_3 and H_2O (300–550°C), calcite decomposes into CaO and CO_2 (600–800°C) and at a temperature of around 800–900°C there can be emergence of tricalcium aluminate($Ca_3Al_2O_6$) and gehlenite ($Ca_2Al_2SiO_7$) (Wu and Liu, 2012). The improvement in the crystallinities of the primary phases causes the particle size and strength to continuously increase during the heat treatment from 150 to 1350°C, as seen from physical property testing and SEM examination. The entire set of findings will serve as a crucial foundation for future research on the thorough use of RM. RM can be a viable raw material for the direct manufacture of iron since it has an iron oxide content that is preferably high (30–60 wt.%). The preparation of glass balls can be done using RM vitrified through melting and granulating with the inclusion of gold tailing and waste limestone. Granulated glass balls demonstrated exceptional resistance to the leaching of heavy metal ions, allowing the RM to be used in glass ceramics and construction materials (Park and Park, 2017). Mombelli et al. (2019) investigated a novel method for producing iron sponge (below 1300°C) or cast iron (above 1300°C) by calcining mixtures of RM-reducing agents, such as pure graphite and blast boiler sludges with various equivalent carbon concentrations. This way of valuing RM can be viewed as an economically viable one because it yields high-quality cast iron that can be used as a raw material in oxygen converters or electric furnaces (Mombelli et al., 2019).

4. FUTURE RESEARCH SCOPES

i. A suitable remediation technology for recycling of RM, which is taken into consideration of the radioactive hazardousness in addition to alkalinity, dispersive nature, potentially toxic elements, etc., should be developed.

ii. Creation of advanced and practical instruments for measuring the effectiveness of the RM remediation process.

iii. Design of mathematical models to determine the optimum levels of RM to be added for various remediation techniques.

iv. Performance of remediated RM in terms of long-term stability and durability need to explored.

v. pH rebound studies of remediated RM should be studied extensively.

vi. Geotechnical and geo-environmental effects due to aging of RM remediated using various technologies need to be investigated

5. CONCLUSIONS

Even if there are a significant amount of accumulated waste products, there is no perfect way to eliminate or clean them up. In deciding between ecology and economy, one should choose ecology to protect future generations and prevent creating more waste simultaneously. Zero waste technology should be our ultimate objective. This can be done only by enhancing the value of RM, so that it can be a raw material for numerous applications, thereby put a relief on natural resources. More aluminum plants could be located in more locations if the RM generation's technology hurdle is resolved. This will produce infrastructure that can be used in different methods to improve the local economy. Site-specific cleanup technologies should be used as a response. The critical review conducted here makes it clear that a single technology cannot address all of RM's negative effects at once. As a result, an integrated remediation strategy that takes into account efficiency, cost effectiveness, risks, resource availability, and time consumption should be developed.

REFERENCES

[1] T. Alshaal, et al., "Phytoremediation of bauxite-derived RM by giant reed," *Environ Chem Lett*, vol. 11, no. 3, pp. 295–302, 2013, doi: 10.1007/s10311-013-0406-6.

[2] V. R. Chate, R. M. Kulkarni, V. M. Desai, & P. B. Kunkangar, "Seawater-washed activated bauxite residue for fluoride removal: waste utilization technique," *J Environ Eng*, vol. 144, no. 5, p. 04018031-1, 2018.

[3]   K. Deelwal, K. Dharavath, & M. Kulshreshtha, "Evaluation of characteristic properties of red mud for possible use as a geotechnical material in civil construction," *Int J Adv Eng Technol*, vol. 7, no. 3, p. 1053, 2014.

[4]   M. Gautam, & M. Agrawal, "Phytoremediation of metals using vetiver (*Chrysopogon zizanioides* (L.) Roberty) grown under different levels of RM in sludge amended soil," *J Geochemical Explor*, vol. 182, pp. 218–227, 2017.

[5]   E. Kalkan, "Utilization of red mud as a stabilization material for the preparation of clay liners," *Eng Geol*, vol. 87, nos. 3–4, pp. 220–229, 2006.

[6]   S. S. Kushwaha, D. Kishan, M. S. Chauhan, & S. Khetawath, "Stabilization of RM using eko soil enzyme for highway embankment," *Mater Today Proc*, vol. 5, no. 9, pp. 20500–20512, 2018.

[7]   P. M. Lavanya, & S. Kumar, "Characterization of RM as a construction material using bioremediation," *Int J Res Sci Adv Eng*, vol. 2, no. 18, pp. 132–139, 2017.

[8]   P. Liu, G. Shao, & R. Huang, "Treatment of Bayer-process RM through microbially induced carbonate precipitation," *J Mater Civ Eng*, vol. 33, no. 5, p. 04021067, 2021.

[9]   W. M. Mayes, I. T. Burke, H. I. Gomes, Á. D. Anton, M. Molnár, V. Feigl, & É. Ujaczki, "Advances in understanding environmental risks of red mud after the Ajka spill, Hungary," *J Sustain Metal*, vol. 2, no. 4, pp. 332–343, 2016.

[10]   T. Mishra, & V. C. Pandey, *Phytoremediation of RM Deposits Through Natural Succession*. Elsevier Inc., 2018.

[11]   M. C. Mishra, N. G. Reddy, & B. H. Rao, "Potential of citric acid for treatment of extremely alkaline bauxite residue: effect on geotechnical and geoenvironmental properties," *J Hazard Toxic Radioact Waste*, vol. 24, no. 4, p. 04020047, 2020.

[12]   D. Mombelli, S. Barella, A. Gruttadauria, & C. Mapelli, "Iron recovery from bauxite tailings red mud by thermal reduction with blast furnace sludge," *Appl Sci*, vol. 9, no. 22, p. 4902, 2019.

[13]   E. Mukiza, L. Zhang, X. Liu, & N. Zhang, "Utilization of red mud in road base and subgrade materials: a review," *Resour Conserv Recycl*, vol. 141, pp. 187–199, 2019.

[14]   H. R. Nikraz, A. J. Bodley, D. J. Cooling, P. Y. L. Kong, & M. Soomro, "Comparison of physical properties between treated and untreated bauxite residue mud," *J Mater Civ Eng*, vol. 19, no. 1, pp. 2–9, 2007.

[15]   I. Panda, S. Jain, S. K. Das, & R. Jayabalan, "Characterization of RM as a structural fill and embankment material using bioremediation," *Int Biodeterior Biodegrad*, vol. 119, pp. 368–376, 2017.

[16]   H. S. Park, & J. H. Park, "Vitrification of red mud with mine wastes through melting and granulation process—preparation of glass ball," *J Non-Cryst Solids*, vol. 475, pp. 129–135, 2017.

[17]   M. Putrevu, J. S. Thiyagarajan, D. Pasla, K. I. S. A. Kabeer, & K. Bisht, "Valorization of RM waste for cleaner production of construction materials," *J Hazard Toxic Radioact Waste*, vol. 25, no. 4, pp. 1–16, 2021.

[18]   Y. Qu, B. Lian, B. Mo, & C. Liu, "Bioleaching of heavy metals from RM using Aspergillus niger," *Hydrometallurgy*, vol. 136, pp. 71–77, 2013.

[19]   S. Rai, S. Bahadure, M. J. Chaddha, & A. Agnihotri, "A way forward in waste management of red mud/bauxite residue in building and construction industry," *Trans Indian Natl Acad Eng*, vol. 5, no. 3, pp. 437–448, 2020.

[20]   S. Rai, K. Wasewar, J. Mukhopadhyay, C. Yoo, & H. Uslu, "Neutralization and utilization of RM for its better waste management," *Arch Environ Sci*, vol. 6, pp. 5410–5430, 2012.

[21]   "RM Project—site dedicated on the valorisation and best practices on bauxite residue." https://redmud.org/ (accessed July 17, 2021).

[22]   N. G. Reddy, B. H. Rao, & K. R. Reddy, "Biopolymer amendment for mitigating dispersive characteristics of RM waste," *Geotech Lett*, vol. 8, no. 3, pp. 201–207, 2018.

[23] P. Sahoo, & J. Joseph, "Radioactive hazards in utilization of industrial by-products: comprehensive review," *J Hazard Toxic Radioact Waste*, vol. 25, no. 3, p. 03121001, 2021.

[24] P. Satayanarayana, G. P. Naidu, S. Adiseshu, & C. V. Hanumanth Rao, "Characterization of lime stabilized redmud mix for feasibility in road construction," *Int J Eng Res Dev*, vol. 3, no. 7, pp. 20–26, 2012.

[25] A. J. Sipayung, D. T. Suryaningtyas, & B. Sumawinata, "Electrokinetic of red mud," *IOP Conf Series: Earth Environ Sci*, vol. 393, no. 1, p. 012090, 2017. IOP Publishing.

[26] C. S. Wu, & D. Y. Liu, "Mineral phase and physical properties of red mud calcined at different temperatures," *J Nanomater*, vol. 2012, pp. 1–6, 2012.

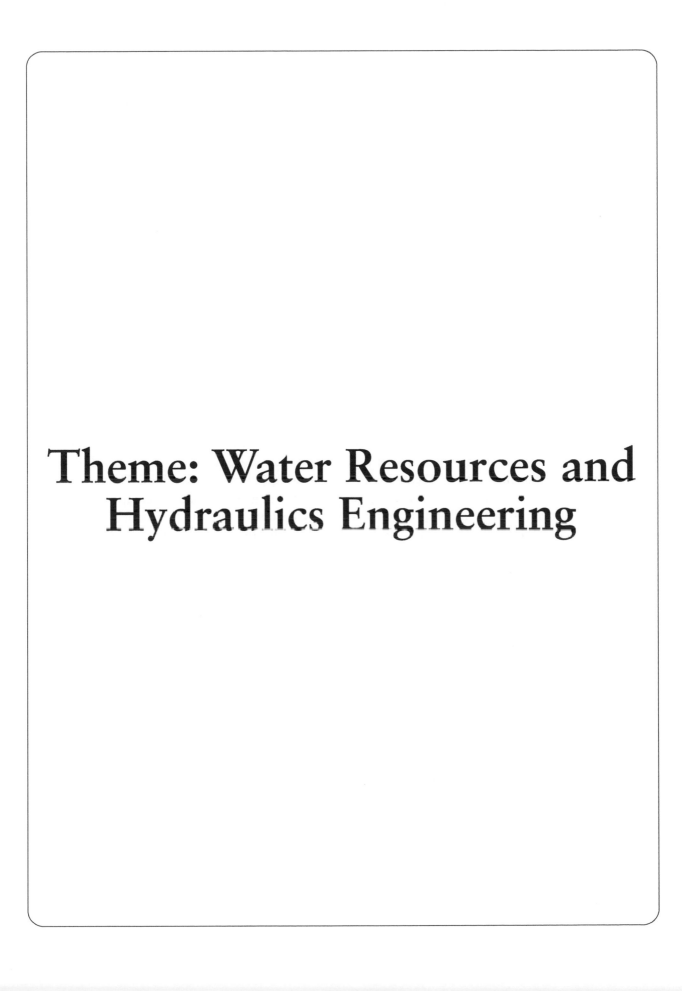

Theme: Water Resources and Hydraulics Engineering

24. Study of Treated Wastewater Using Phytorid Technology in RIT Campus

Achal Chand K [1], Amal Z [1,*], Muhammed Ershad P [1], Vishnu B [1] and Rejoice A A [2,*]

[1]Rajiv Gandhi Institute of Technology, Kottayam, India

[2]Department of Civil Engineering, Rajiv Gandhi Institute of Technology, APJ Abdul Kalam Technological University, Kottayam, India

*Corresponding authors: 18bc12289@rit.ac.in, rejoiceabraham@rit.ac.in

ABSTRACT: Earth is often referred to as the blue planet as oceans and other water bodies occupy around three-fourths of the Earth's surface and constitute 70% of the Earth. But only about 1% is available as freshwater. Water is essential for the existence of life and our bodies are mostly made of it. Due to rapid growth in population and industrialization, water consumption has increased, leading to water scarcity and related water pollution problems. The overall use of water is going to increase manyfold in the years to come and the effects of climate change are going to become more and more evident than ever before. It is predicted that water stress will get even worse. Greywater (GW) is a promising candidate as a substitute water source in semi-arid regions of the globe. Although one best way to treat GW cannot be said, artificial wetlands have proven to be an efficient and cost-effective option. The study is based on an environmentally friendly approach to water treatment on wastewater produced on the college campus. The work was done in the Rajiv Gandhi Institute of Technology cafeteria using phytorid technology. The fact that distinct phytorid technology from conventional methods is that it uses natural components like plants, soil, gravel, stone, charcoal, etc. for treatment. *Heliconia psittacorum* (Parrot Heliconia), *Chrysopogon zizanioides* (Vetiver), *Colocasia esculenta* (Green Taro), *Canna indica* (Indian Shot), which are indigenous to the region and abundantly available in the Western Ghats, were used in the constructed wetland. *The water quality assessment parameters like temperature, pH, turbidity, hardness, chloride concentration, dissolved oxygen, phosphate, BOD, and COD are used to validate the performance of the proposed greywater treatment system. The test was conducted to confirm the efficiency of the treatment in the water sample, obtaining very high removal yields on turbidity (97.42%), total suspended solids (89.82%), BOD (89.32%), and COD (93%).*

KEYWORDS: Wastewater Treatment, Phytorid Technology, Water Quality Parameters, Domestic Wastewater, Constructed Wetland

1. INTRODUCTION

Untreated sewage is sometimes dumped on agricultural lands as a way of sewage land disposal, particularly in underdeveloped nations. However, this technique of disposal raises public concern because of the high likelihood of phytotoxicity and sizable chances of metal ions being incorporated into food grains, which in turn could affect the food chain. There is a huge possibility for damage to groundwater because of the leaching of excess nitrogen and phosphorus in effluents due to ongoing sewage effluent use (Witthayaphirom et al., 2020).

The National Environmental Engineering Research Institute (NEERI), Nagpur, has formed a ground-breaking technology that is based on the natural process of sewage treatment using a system

Chapter 24 DOI- 10.1201/9781032657271-24

of built wetlands. This technology is called phytorid technology (Collivignarelli et al., 2020). Phytorid technology is developed in such a manner that it can be easily utilized in cities along with rural areas for wastewater treatment with much focus on the economy and can be used for a wide variety of applications (Bangar et al., 2019). The technology uses certain plants with phytoremediation capability, usually found in a wetland (Al-Ajalin et al., 2020).

2. LITERATURE REVIEW

2.1. Phytorid technology

By making use of certain varieties of plants which do not essentially require soil for growth and can take nutrients directly from the wastewater, a self-sustainable technology for treating wastewater is developed which works on the principle of natural wetlands and has been named phytorid technology (Sawant et al., 2020). These plants assist as sinkers and movers of nutrients throughout. The technology is protected by patents in Europe, Australia, and India. This technique has been approved by the Council for Scientific and Industrial Research and the NEERI (National Environmental Engineering Research Institute; Aalam and Khalil, 2019). It is an improved wetland ecosystem for the treatment of wastewater. It involves the proper utilization of biological treatment capacity with optimized engineering parameters. Aquatic or semi-aquatic plants are planted in the filterable wetland. Wastewater will enter the unit through vertical and horizontal units that are constructed for better hydraulics and for providing sufficient retention time.

2.2. Performance of the treatment

In order to quantify the performance of the treatment, a series of experiments were completed with various parameters and analyzed the ability of the process to change the features of the water in response to their reaction (Bhamare et al., 2020). The percentage removal of the parameters clearly shows the performance of the sewage water treatment plant. The important parameters and percentage removal of pollutants in an STP using phytorid technology are shown in Table 1.

Table 1. Performance of phytorid for urban waste pollutant (Bhamare et al., 2020).

Performance (% removal)	Parameter
90–95	Biochemical oxygen reaction
85–95	Chemical oxygen demand
90–95	Total suspended solids
60–85	Total nitrogen
50–80	Phosphate
80–90	Turbidity
70–85	pH

3. METHODOLOGY

The study area is located at the main cafeteria of Rajiv Gandhi Institute of Technology, Kottayam, situated in the southern region of Kerala, India (9° 34′ 37.416″ N, 76° 37′ 21.72″ E). The study area receives an average annual rainfall of 3130.33 mm, and the average annual temperature is around 27°C. Also, the study area shows a rather consistent climate throughout the year as it is located in a semiarid region. The major components of phytorid technology are sewage collection tanks, phytorid beds, and treated water storage.

3.1. Design of phytorid tank

3.1.1. Total water quantity

The design of constructed wetland unit for wastewater treatment was by analyzing the volume of wastewater to be treated. The volume of wastewater generated was calculated using the statistical data about the number of students and faculty using the cafeteria and the net quantity of water required by an individual.

- Population using the cafeteria = 400 individuals
- Water supply requirement = 0.6 liters per capita per day
- Assuming the net quantity of sewage produced is equal to 90% of the accounted water supplied from the waterworks.
- Net quantity of sewage produced = 400 × 0.6 × 0.90 = 216 L = 250 L/day = 0.25 m³/day
- The quantity of wastewater produced is enough to support the minimum quantity of water required to maintain the constructed wetland condition.

3.1.2. Dimensions of phytorid tank

The sewage volume is mainly the important consideration for sizing constructed wetland units which usually have a length-to-width ratio of 2:1. The depth of the tank is capped at a maximum of 60 cm. This maximum depth of 60 cm is provided to ensure that there is adequate water available in the region to sustain an aerobic environment throughout the phytorid treatment process.

- Net quantity of sewage produced = 0.25 m³/day
- Proving an extra of 20 L/day, that is, 0.02 m³/day
- Gross quantity of sewage = 0.25 m³/day + 0.02 m³/day = 0.27 m³/day
- Assuming the porosity of the phytorid bed as 0.45
- $Porosity, \emptyset = \frac{V_v}{V}$
- V_v is the volume of voids = 0.27 m³
- V is the total volume of the tank
- $0.45 = \frac{0.27}{V}$
- Thus, the total volume of the tank $= \frac{0.27}{0.45} = 0.6\ m^3$
- Since the depth of the phytorid tank is capped at a maximum of 0.60 m
- Surface area = Assuming an $L{:}B$ ratio of 2:1, L is the length of the phytorid tank and B is the width of the phytorid tank
- $\frac{L}{B} = \frac{2}{1} \equiv L = 2B$
- Surface Area $= L \times B = 2B \times B$
- $B = \sqrt{\frac{1}{2}} = 0.7m$
- $L = 2B = 2 \times 0.7 = 1.4$ m
- Providing 10 cm extra on the length of the phytorid tank for the inlet and outlet,
- $L = 1.4 + 0.1 = 1.5$ m
- Providing 20 cm extra on width of phytorid tank, $B = 0.7 + 0.2 = 0.9$ m
- Providing a freeboard of 20 cm for the depth, $D = 0.6 + 0.2 = 0.8$ m
- Total volume of phytorid tank $- L \times B \times H = 1.08$ m³
- Thus, the overall dimensions of the tank are as follows,
  Length = 1.5 m   Width = 0.9 m   Depth = 0.8 m

3.1.3. Design of baffle walls

- Length of individual compartment $= \frac{L}{4} = \frac{1.5}{4} = 37.5$ cm
- Number of compartments $= \frac{1.5}{0.375} = 4$ compartments
- Number of baffle walls = No. compartments - 1 = 4 - 1 = 3 number of baffle walls
- Height of end baffle walls = 0.3 m
- Height of intermediate baffle walls = 0.4 m

Thus three baffle walls are provided at 37.5 cm spacing and the end baffle walls have a height of 30 cm and the intermediate baffle wall has a height of 40 cm. Figure 1 shows the digital model of the phytorid tank.

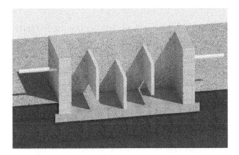

Figure 1. *Section of the phytorid tank.*

3.2. Design of phytorid bed

Wastewater treatment in a phytorid bed takes place by means of natural operations and processes such as filtration, sedimentation, nutrient uptake by plants and microbial action in a constructed system which is filled with media such as gravel. Certain plant species which are known for their good nutrient intake efficiency are planted in the phytorid bed.

3.2.1. Layers of phytorid bed

The gradual reduction in particle size of filler materials helps in the filtration of the wastewater and also aids in purification.

- Layer 1: 20 cm thick: coarse aggregate of 40 mm effective diameter
- Layer 2: 35 cm thick: coarse aggregate of 20 mm effective diameter
- Layer 3: 30 cm thick: coarse aggregate of 10 mm effective diameter
- Layer 5: 35 cm thick: coarse aggregate of 8 mm effective diameter
- Layer 6: 10 cm thick: activated carbon
- Layer 7: 30 cm thick: coarse aggregate of 4.75 mm effective diameter
- Layer 8: 80 cm thick: fine aggregate
- Layer 9: 15 cm thick: soil for plant growth

Some intermediate layers of broken bricks and stone chippings are also added in between. Alum is also provided occasionally for better efficiency.

3.3 Construction

The phytorid tank is constructed in the field near the cafeteria as per the dimensions and specifications provided in the design. The finalized dimensions of the tank are such that it has a length of 1.5 m, a width of 0.9 m, and a net height of 0.8 m. The schedule of work is as follows: site preparation, setting out of the tank, masonry works, concrete works, attaching of baffle walls, filling of the tank with materials, and collection tank set up.

Figure 2. *Completed phytorid tank.*

4. RESULTS AND DISCUSSION

The collected wastewater was subjected to several tests to determine the wastewater characteristics before and after treatment and to determine the removal efficiency of the phytorid bed constructed. Table 2 shows the different water quality parameters and their percentage removal efficiencies.

Table 2. Comparison of the efficiency of different parameters.

Sl. no.	Parameter	Initial value	Final value	Removal efficiency (%)	Remarks
1.	pH	8.66	7.51	–	Permissible (6.5–8.5 as per B.I.S.)
2.	Turbidity	27.1	0.7	97.42	Permissible (5 NTU as per B.I.S.)
3.	Hardness	243.33	163.33	32.87	Permissible (300 mg/L as per B.I.S.)
4.	Chloride	187.44	166.61	11.11	Permissible for irrigation (250 mg/L as per B.I.S.)
5.	D.O.	1	2.74	–	Permissible (>6 mg/L as per B.I.S.)
6.	Phosphate	2.2	0.763	65.50	Permissible (5 mg/L as per Environment Protection Rule)
7.	T.S.S.	570	58	89.82	Permissible (100 mg/L as per Environmental Standards)
8.	BOD	152	16.23	89.32	Permissible (30 mg/L as per Environmental Standards)
9.	COD	524	36.68	93	Permissible (250 mg/L as per Environmental Standards)

- The water quality parameters were identified by testing and accordingly, the layers of the filter media were fixed. The parameters also shed light on the plants to be used.
- Phytorid plants like *Heliconia psittacorum* (Parrot Heliconia), *Chrysopogon zizanioides* (Vetiver), *Colocasia esculenta* (Green Taro), *Canna indica* (Indian Shot), which are indigenous to the region and abundantly available are used for the constructed wetland to facilitate phytoremediation.
- The effluent from the phytorid tank was tested for the same water quality parameters as before and by comparing the initial and final values the removal efficiency was calculated. The above-obtained results (Table 2) provide a respectable view of the effectiveness of the phytorid technology.

5. CONCLUSIONS

Phytorid technology was used to facilitate the treatment of the wastewater from the cafeteria and to make it available for other secondary purposes like gardening and toilet flushing, a phytorid treatment plant has been successfully designed and implemented.

- The designed dimensions of the tank are as follows,
- Length = 1.5 m Width = 0.9 m   Depth = 0.8 m

- The design also provided the use of three number of baffle walls, end ones and intermittent ones of height 30 cm and 40 cm, respectively.
- The use of baffle walls significantly reduces the length of the tank required.

Phytorid technology can be best utilized for the efficient treatment of wastewater for institutions, and restaurants with minimal energy requirements and high efficiency, especially in developing countries of the world.

REFERENCES

[1] Aalam, T., and Khalil, N., "Performance of horizontal subsurface flow Constructed Wetlands with different flow patterns using dual media low strength municipal wastewater: a case of pilot scale experiment in a tropical climate region", *J Environ Sci Health*, vol. 1, pp. 1093–4529, 2019.

[2] Al-Ajalin, F. A. H., Idris, M., Abdullah, S. R. S., Kurniawan, S. D., and Imron, M. F., "Effect of wastewater depth to the performance of short-term batching experiments horizontal flow constructed wetland system in treating domestic wastewater", *Environ Technol Innov*, vol. 20, pp. 101–106, 2020.

[3] Bangar, A. T., Kavedi, O. K., Yusuf, K. M., Patil, D. A., and Sharan, R., "Treatment of wastewater by process of Phytoremediation using Sunflower plant", *Int Res J Eng Technol*, vol. 6, no.4, pp. 2–8, 2019.

[4] Bhamare, S., Patil, R., Kolate, V., Chaudhari, S., Gope, P., Mehar, A., Darge, A., and Bhagat, S. R., "Analysis and design of sewage treatment plant using Phytorid Technology", *Int J Res Appl Sci Eng Technol*, vol. 8, no. 6, pp. 2321–9653, 2020.

[5] Collivignarelli, M. C., Miino, M. C., Gomez, F. H., Torretta, V., Rada, E. C., and Sorlini, S., "Horizontal flow constructed wetland for greywater treatment and reuse: an experimental case", *Environ Res Publ Health*, vol. 17, pp. 2317, 2020.

[6] Sawant, M., Aware, K., and Honrao, Y., "Phytorid technology", *Int J Sci Res Eng Trends,* vol. 6, no. 5, pp. 2797, 2020.

[7] Witthayaphirom, C., Chiemchaisri, C., Chiemchaisri, W., Ogata, Y., Ebie, Y., and Ishigaki, T., "Organic micro-pollutant removals from landfill leachate in horizontal subsurface flow constructed wetland operated in the Tropical climate", *J Water Process Eng*, vol. 38, pp. 101581, 2020.

25. Spatio-Temporal Variations of Hydrologic Components

Akshay Ranjith[*]

Department of Civil Engineering, Indian Institute of Science, Bangalore, Karnataka, India

[*]Corresponding author: rakshay@iisc.ac.in

ABSTRACT: One of the most important natural resources on Earth is water. Water, in all of its forms, is an essential element that propels the functioning of the Earth as a whole system. To comprehend how water influences the system, researchers study water in all of its forms. A precise and representative form of water, such as precipitation, stormwater flows, evaporation, etc., is needed for many aspects of development, including urban development, building projects, water distribution, treatment of sewage, etc. However, it might be time-consuming to gather the many hydrological components' quickly fluctuating temporal and geographical distribution. The study aims to understand and analyze various hydrologic components and their variations in the temporal domain. The Model Parameter Estimation Project, abbreviated as MOPEX of the Global Energy and Water Exchange Hydrometeorology Panel, which offers top-notch hydrological and meteorological data along with a wide range of attributes like soil characteristics, climate normals, land use, etc., is the basis of the open-source dataset used in this study. To determine several components of surface water hydrology, including precipitation, streamflow, runoff, and changes in storage, this information is carefully employed. These findings are then examined from temporal patterns, which include seasonal patterns, variations across months, trends, and modality. The correlation between annual precipitation and streamflow is also examined. Software tools like R Studio for outlier and frequency analysis, MATLAB and MS Office for time scale changes, plotting temporal variations, and trend analysis were used for this study. The study provides a pathway to analyzing and understanding a dataset of hydro-meteorological variables to bring out a picture of the various hydrological components which are weaved into the lives of the habitants of this planet Earth.

KEYWORDS: MOPEX, Precipitation, Runoff, Spatio-Temporal Variation, Streamflow

1. INTRODUCTION

1.1. General

Water is an inorganic, flavorless, odorless, transparent, and almost colorless chemical substance that makes up the majority of the hydrosphere of the planet Earth and the fluids of every living thing that exists. Water with all this immense importance has led to the development of human civilization with various spheres falling under as agriculture, industrial, domestic, hydropower generation, for navigation and recreation. The list seems endless. For sustainable development, forecasting and planning lies in as the keystone. This is where the various spheres of understanding the water as a holistic system come into play. The water resources division puts in the efforts to understand and use the water resources in a sustainable manner. Hence, at the fundamental level, lies the importance of understanding the different components that drive the water all around the planet. The various forms of water that are of primary importance are looked into in this study, namely, the precipitation, streamflow, runoff, and changes in storage. These are important when it comes to planning and implementing any project in direction of holistic development as water is a fundamental part of anything in this planet, which stands proven as water can be connected or traced into anything in this planet Earth.

Chapter 25 DOI- 10.1201/9781032657271-25

1.2. Study area

The study areas consist of three different places based on the datasets obtained from MOPEX. Owing to the abundance of data of excellent quality with the minimal aberrations available in these locations, they were selected as the research areas. The study areas chosen are Mercer, Turner Center, Yarmouth, and Conway in Maine and Plymouth in North Hampshire, United States of America. The stream flow measurements are collected from a stream gauge installed on the Sandy, Nezinscot, and Pemigewasset rivers, which have watershed areas of 1337, 438, and 1611 km^2, respectively, and the limits of which are depicted in Figure 1. The precipitation metrics were collected via the United States Geological Survey (USGS) stations (hydrologic unit: 1030003, 1040002, and 1070001) located at Mercer, Turner Center, and Plymouth, respectively. The need for understanding water systems has become an essential part of human development. The different components of water mainly precipitation, streamflow, runoff, and changes in storage is to be measured accurately and studied to understand the trend and its various correlations. These are useful some of the primary steps toward counteracting measures for the aggravating climatic extremes. They can be used to forecast and predict these components for the design year.

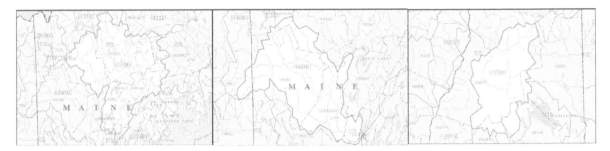

Figure 1. Location of sites 1, 2, and 3 (numbered in clockwise direction). Courtesy of USGS.

2. METHODOLOGY

2.1. General

The study involved analysis and understanding different components of hydrological cycle whereby inferences about spatio-temporal variations are studied. On the basis of literature reviews, the following goals were developed; to assess and draw inferences from monthly and annual precipitation, to assess and draw inferences from monthly and annual streamflow, to estimate and draw inferences using precipitation-streamflow relationships and to estimate and draw inferences about change in annual storage.

2.2. Collection and processing of data

2.2.1. General

In order to obtain the following objectives, we need daily precipitation, streamflow, and evaporation for a particular location which should be reliable. Since Global Energy and Water Exchange Hydrometeorology Panel offers hydrological and meteorological details along with a wide range of characteristics like soil characteristics, climate normals, land use, etc., we use the historical data of the location of interest that has been collected from Model Parameter Estimation Project.

2.2.2. MOPEX dataset

The Model Parameter Estimation Project (MOPEX), run by the GEWEX (Global Energy and Water Exchange) Hydrometeorology Panel (GHP), provides the historical data for the target site. The National Oceanic and Atmospheric Administration's website can be used to get the data.

Each day's data is provided for each of the three datasets from 1948 to 2003. To eliminate outliers, the dataset is analyzed in R Studio. For huge number of missing data for a particular month or year, that entire month or year is removed, while if the missing data are very less in their frequency, the same are filled using Effron bootstrap method using generation of uniformly distributed random numbers. The climatic potential evaporation mentioned above is determined using the NOAA Freewater Evaporation Atlas (Farnsworth et al., 1982). To obtain the actual evaporation, the provided potential evaporation is transformed using the Schreiber method for calculating actual evaporation:

$$\frac{AE}{P} = 1 - e^{-\frac{PE}{P}} \tag{1}$$

At the annual time scale, the variables are defined as follows: "P" represents annual precipitation, "E" represents actual evaporation, and "PE" represents potential evaporation. To obtain data at both monthly and yearly scales, the provided daily values are transformed into monthly and yearly values using MATLAB's time series analysis tool. This process is applied to the given daily data for precipitation, streamflow, and potential evaporation. It is worth noting that all the quantities, including precipitation, streamflow, and evaporation, are measured in millimeters of water.

2.3. Temporal variations of precipitation and streamflow

The temporal variations precipitation and streamflow are plotted using MATLAB as well as MS Excel. MATLAB uses plot function along with the add-ons like title of the figure, axis title, font size, etc. Similarly, MS Excel is used to plot the temporal variations of the same. A linear trend as well as a two period moving average trend is fitted to the temporal variations using MS Excel functions.

2.4. Annual storage change

The processed dataset contains precipitation, streamflow, and potential evaporation in annual scales. The actual evaporation is obtained using the annual precipitation and annual potential evaporation using the Schreiber formulation of actual evaporation as shown in Equation (1). The annual storage change is calculated using the above hydrologic components as shown in Equation 2.

$$\textit{Annual storage change}, \Delta S = P - R - AE \tag{2}$$

where all the variables are at annual time scale, P is the annual precipitation, R is the annual streamflow and AE is the actual evaporation. The temporal variation of annual storage changes is plotted using MS Excel and it is analyzed to understand the deficit and surplus years. For all the above hydrologic components, they are studied for three datasets which are from three different locations as shown in Section 1.2, this helps us to understand the spatial variation of the hydrologic components considered in this study.

3. RESULTS AND DISCUSSION

3.1. General

The acquired MOPEX datasets which contain the data for the target sites were collected and outlier study was conducted. The datasets are then converted to different time scales using MATLAB. The various hydrologic components are calculated and analyzed. The spatio-temporal variations are plotted for each of the hydrologic component and for each of the study area using MATLAB as well as MS Excel. Each of the hydrologic components are studied and the valuable information obtained are as follows for each of the study area.

3.2. Collection and processing of data

For dataset 1, there are no outliers in case of precipitation and potential evaporation, but 3281 (approximately 9 years data—from 1979 to 1987 and 2002 and 2003) streamflow values are missing. For dataset 2, There are no outliers in case of precipitation and potential evaporation, but 2283 (approximately 7 years data—from 1996 to 2003) streamflow values are missing. For dataset 3, no data is available in case of potential evaporation, there are no outliers in case of precipitation, but 1096 (approximately 3 years data—from 1948, 1949, 2002, and 2003) streamflow values are missing. For dataset 4, there are no outliers in case of precipitation and potential evaporation, but 457 (approximately 2 years data—from 2002 to 2003) streamflow values are missing. For dataset 5, there are 11 missing values in the year 1948 for precipitation, which are filled in using Effron bootstrapping. There are no outliers in case of potential evaporation, but 457 (approximately 2 years data—from 2002 to 2003) streamflow values are missing. Hence, any calculations involving streamflow, the missing data is not considered.

3.3. Temporal variations of precipitation and streamflow

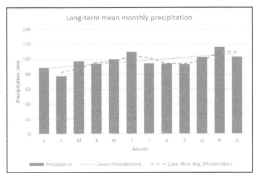
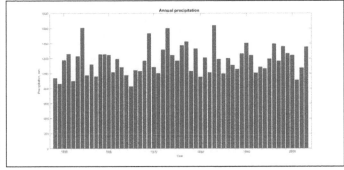
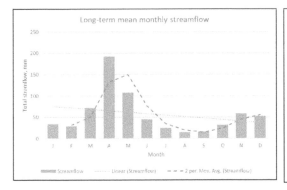
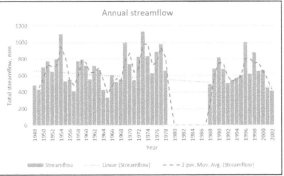
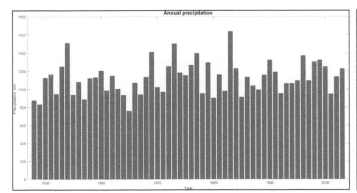
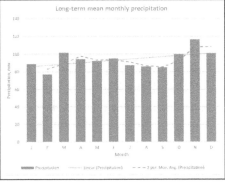

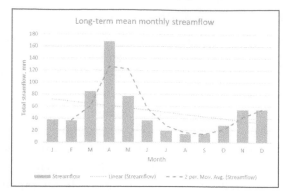

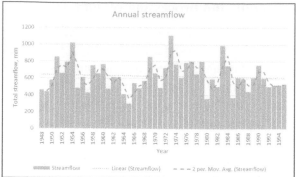

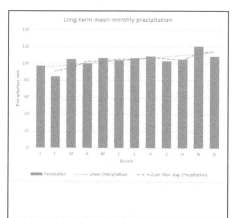

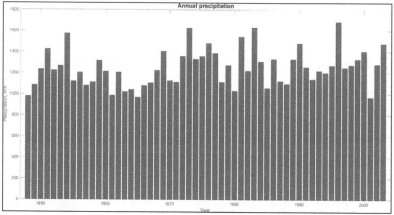

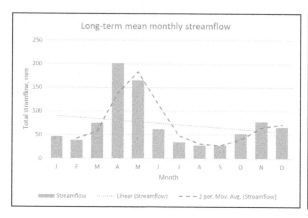

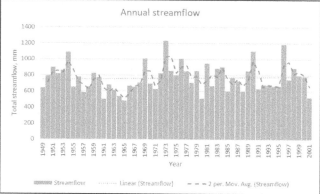

Figure 2. *Various precipitation and streamflow in different timescales for datasets 1, 2, and 3.*

The precipitation is observed to be less variable across the months for all three datasets. There is a very minute presence of seasonal pattern as the precipitation seems to be increasing and reaches the local maxima during the months of June and November. The two-period moving average model shows a cyclic trend while the linear trend shows a continuously increasing trend, which indicates that the precipitation increases to toward the end of the year. No significant trend in annual precipitation is observed, but slight higher peaks are seen to be evident during every 10–20 years approximately with the second one being a very low flow rather than a peak. The monthly and the annual temporal variation of streamflow are shown in Figure 2.

The streamflow is observed to be highly variable across the months with maxima in April for all the study areas. Dominant seasonal pattern can be observed with a bimodality in seasonality of streamflow. There is a significant difference in the wet flow and dry flows of about 178, 154, and 173 mm of mean monthly streamflow for study areas one, two, and three, respectively. The two-period moving average model shows a cyclic trend with a flattened trough, which indicates that the dry months are comparatively of higher duration; while the linear trend shows a continuously decreasing trend, as the dry flows are of longer duration. No significant trend in annual streamflow is observed, but two period moving average model suggests that higher peaks are seen to be evident during every 4–6 years approximately with a slight decreasing trend in peaks. The linear trend shows a decreasing trend in streamflow, which could be due to the aggravating climate change scenario which predicts an increasing water shortage.

3.4. Annual storage change

In all the figures which show the annual storage change, the red color bars indicate water deficit years whereas blue color bars indicate water surplus years. The temporal variations of annual storage change are plotted as shown in Figure 3. It is evident from the figure that there are more water deficit years than water surplus years, which implies that all the three study areas could have faced shortage of water during the periods of high deficit years. The gaps are due to missing data. Also, it is evident that there are very few surplus years during the period of 1948–1964 for all the three datasets.

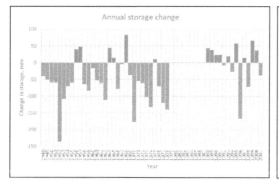
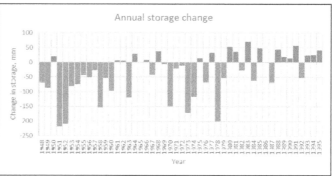
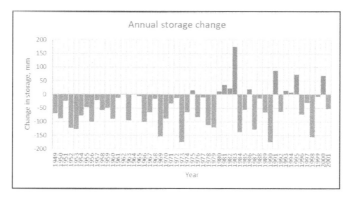

Figure 3. Annual storage change for datasets 1, 2, and 3.

3.5. Spatial variation of the hydrologic components

Comparing the various datasets which are corresponding to the respective study areas as mentioned in Section 1.2. It is evident from the above figures that for precipitation and streamflow are very similar. This also verifies the fact that the study areas are close by and could possibly lie in the same climate regime. The precipitation–streamflow relationships are similar with the runoff ratios lying in the range of 56% to 62% and all of them suggest a very strong correlation. The annual storage changes are also similar with very high water deficit years in the period of 1948–1960. In all the cases except for dataset 3 where the data is insufficient, the water deficit years are more than water surplus years.

4. CONCLUSION

This study analyzes the spatio-temporal variations of hydrologic components. The spatial variations are analyzed using datasets from five different study areas while the temporal variations are analyzed at monthly as well as yearly scales. The study reveals the following details: there is a strong seasonal trend in the case of streamflow, while the precipitation does not show any strong seasonal trend at all the study areas. Long-term monthly precipitation at all the study areas has the maxima during the month of November, which is the winter season at the onset of rainfall. Long-term monthly streamflow at all the study areas have their maxima during the month of April with the minima during the month of August and September. High streamflows are observed every 4–6 years at all the study areas. The streamflow in study areas 1, 2, and 3 decreases while increases for study areas 6 and 5, which could be an indicator of the climate change as for the dynamics in higher latitudes (for study area 1, 2, and 3) seems to decrease the streamflow while the opposite happens for study area in relatively lower latitudes. The precipitation–streamflow has very strong correlation for all the study areas with the runoff ratios lying in the range of 56–62%. All the study areas have more deficit years with them being concentrated in the period of 1948–1960.

Limitations and scope of the study: The study does not consider the use of complex statistical tools to analyze the historical time series data. One could fit a three or higher parameter models to the data and can be used to predict future flows. The study does not incorporate the effects of climate change. One could use inputs from global climate models to predict future flows. A richer dataset could be considered, which may include evapotranspiration, infiltration and usage of digital elevation models can accurately predict runoffs.

REFERENCES

[1]   Farnsworth, R. K., Thompson, E. S., and Peck, E. L., "Evaporation Atlas for the contiguous 48 United States," NOAA Technical Report, NWS 33, Washington, DC, pp. 24–26, 1982.

26. A Hybrid RF-LSTM Model for Daily Streamflow Prediction of Greater Pamba River Basin, Kerala Incorporating Dominant Hydro-Climatic Drivers

Arathy Nair G R [1,*] and Adarsh S [2]

[1,2]Department of Civil Engineering, TKM College of Engineering Kollam, APJ Abdul Kalam Technological University, Kerala

*Corresponding authors: grarathynair@gmail.com, adarsh_lce@yahoo.co.in

ABSTRACT: Reliable prediction of streamflow is considered a prerequisite for flood risk studies, watershed management, dam, and reservoir construction, and river training works.. In the recent past, there exist different practices for streamflow prediction, ranging from conventional methods of statistical time series modeling to hybrid artificial intelligence modeling. The major concerns in such techniques are the data scarcity and uncertainties bound up with the climatic conditions. Long and continuous time series data is essential to preside an accurate prediction. Climatic changes also induce tremendous impacts on water resources. A hybrid multivariate time series prediction model deployed using deep learning based Long-Short Term Memory (LSTM) technique, along with the Random Forest (RF) feature selector for choosing the most influencing climatic factors, is established in this study. The applicability of the proposed method is demonstrated by considering the daily streamflow, maximum and minimum temperature, precipitation, and root-zone soil moisture data, all lagged data from 1 to 7 days, spanning within a time period of 25 years (1990 to 2015) from three different stations Malakkara, Kalloopara and Thumpamon of Greater Pamba River Basin (GPB) of Kerala, India. The major climatic predictors for each station, determined using RF feature selector, exhibited a significant influence of temperature and precipitation in streamflow prediction. The predicted LSTM model results are evaluated using different criteria: Root Mean Square Error (RMSE), Adjusted R-Squared value (R2), Mean Absolute Error (MAE) and Mean Absolute Percent Error (MAPE). The model demonstrates compelling performance with R2 as 0.972, 0.921, and 0.981, RMSE as 0.283, 0.291, and 0.251, MAE as 0.231, 0.271, 0.177, and MAPE as 9.8%, 10.02% and 6.2% for all the three stations, respectively.

KEYWORDS: Streamflow, Time Series Modeling, Artificial Intelligence, Long-Short Term Memory Technique, Random Forest, Climatic Predictor

1. INTRODUCTION

Today, climate change is a major worry, its effects on the frequency of hydrological extremes such as floods, droughts, heat waves, and cyclones are noteworthy. Among the threats mentioned, flooding-related risks are the most prevalent and widely dispersed. As a result, flood forecasting is always a crucial topic of research. Accurate streamflow prediction is crucial because a timely and exact warning gives plenty of time for more mitigating efforts and less damage from the disaster (Hu et al., 2020).

Chapter 26 DOI- 10.1201/9781032657271-26

Streamflow is regarded as a complicated non-linear time series data that is simultaneously impacted by a number of variables, including precipitation, sustained temperature, and other hydro-climatic forces (Ha et al., 2021; Choi et al., 2021).

In the recent past, there exist various practices for the prediction of streamflow, ranging from conventional methods of statistical time series modeling to hybrid artificial intelligence modeling (Hu et al., 2020; Ha et al., 2021; Yilmaz and Onoz, 2020). Mehdizadeh and Sales (2018) studied the prediction of monthly streamflow of selected basins, by applying Auto-Regressive (AR) and Moving Average (MA) models. Yilmaz and Onoz (2020) made a comparative evaluation of different statistical methods for daily streamflow estimation at the selected basins. Three statistical methods using the smoothed data by moving average technique was considered and they are DAR-MA (Single Donor Drainage Area Ratio), MDAR-MA (Multiple Donor Drainage Area Ratio) and ISW-MA (Inverse Similarity Weighted). Chong et al. (2019) proposed Wavelet Transform (WT) based models for handling time series modeling of daily, monthly, and yearly streamflow. Sahay and Srivastava (2014) discussed the combination of Wavelet Transform and Genetic Algorithm based Neural Network for forecasting monsoon river flows one day ahead. Adnan et al. (2017) investigated the ability of two different soft computing techniques like Artificial Neural Network (ANN) and Support Vector Machine (SVM) for modeling the monthly streamflow data. The model performance evaluators showed that the SVM model outperforms the ANN model in predicting the monthly streamflow. Hassan and Hassan (2021) proposed the methodology for the improvement of ANN based streamflow forecasting models through different techniques of data preprocessing. Recent studies conducted for streamflow prediction utilize the deep learning based Long-Short Term Memory (LSTM) model (Hu et al., 2020). They concluded that the LSTM model for prediction would result in better model stability and reliability, since it is free from conventional issues like exploding and vanishing gradient.

The objectives considered in this study are as follows: (i) to develop LSTM model for streamflow prediction, by considering lagged values of hydro-climatic variables; (ii) to propose a RF based feature selection for most dominant drivers for daily streamflow prediction; (iii) to propose a hybrid RF-LSTM model for streamflow prediction considering most dominant input drivers and compare its performance with standalone LSTM mode.

2. MATERIALS AND METHODS

2.1. Study region and datasets

The Greater Pamba River Basin (GPRB), which consists of three different rivers, Pamba, Achencovil, and Manimala, flowing through Kerala, India; is considered as the study region (Figure 1). The streamflow gauge stations of Thumpamon, Malakkara, and Kalloopparain this basin are considered. The basin is around 4,500 square kilometers in area and is significant while dealing with the hydrological study of Kerala. According to the reports of Central Water Commission, this basin is one of the worst-hit basins of greater floods of 2018, with a rainfall depth of around 538 mm for three consecutive days of 15th to 17th August 2018.

The study was conducted using the streamflow data of the three stations, spanning a time period of 25 years (1990-2015) (collected from WRIS), the precipitation data, the maximum and minimum temperatures, and root-zone soil moisture data (collected from NASA Power website, https://power.larc.nasa.gov) to predict the future streamflow of the same locations.

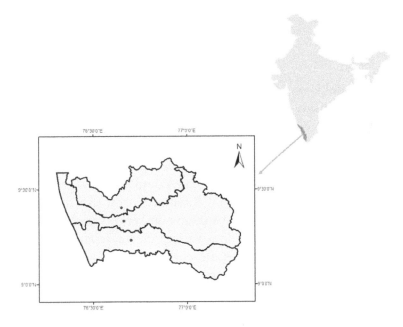

Figure 1. The location of streamflow stations in Greater Pamba River Basin, Kerala, India

2.2. Methods

2.2.1. LSTM network

Recurrent neural networks (RNNs), of which LSTM networks are one type, are characterized by their capacity to remember long-term dependencies while learning the relationships between diverse elements. RNN's limitations include the inability to allocate long-term memory due to its vanishing gradient problem. This is the situation in which LSTM model comes as the savior. The aforementioned vanishing gradient problem can be fully eliminated in LSTM, by making use of an additional forget gate (Figure 2). This simply helps in the long term allocation of the memory.

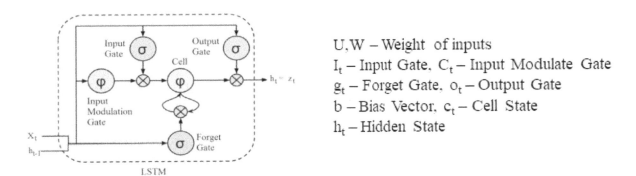

Figure 2. Long- Short Term Memory Cell.

2.2.2 Basic and hybrid LSTM prediction model

In this study, deep learning based LSTM model is used for streamflow prediction. Here, various factors including hydro-climatic parameters that are lagged up to a week are considered. The basic prediction model is fed with all of this data without feeding them into a feature selection module. This may result in over-fitting, which typically occurs when a model is proposed and attempts to incorporate all of the implementation's data points. As a result, the model's efficiency will be lowered. Another approach is suggested by taking into account a module for significant feature selection to address these problems in the prior LSTM model.

2.2.3 *Proposed Methodology*

The main focus of this model is on identifying the variables from the considered dataset that have the greatest influence. This study used cross-correlation analysis to pinpoint the characteristics among the ones described above that have a significant impact on the streamflow prediction Q(t+x).

Second part deals with the feature selection RF module. Feature selection is mandatory while dealing with a large number of feature sets, so as to eliminate comparatively irrevelant variables, improving the prediction model's performance and accuracy. Random Forest (RF) is a handy technique that can solve the selection problem even when there are a large number of variables. This method incorporates feature importance, which can be calculated in a variety of ways. The Gini significance approach is employed in this study. Internal node features are chosen using some criterion, which for selection issues can be gini impurity. The cross-correlation parameters are given into this model to further reduce the dataset and get the optimum feature configuration. Finally, the selected features are fed into the endmost LSTM module for future prediction.

3. RESULTS AND DISCUSSION

In this section, the performance of the basic and hybrid models for streamflow prediction was analyzed. The datasets of streamflow collected for a period of 25 years are divided into training phase (80%) and testing phase (20%), by considering the previous literature (Hu et al., 2020; Ha et al., 2021). The results from the first module of the hybrid RF-LSTM model (i.e., Cross-Correlation phase) carried out, for each of the three stations are shown in Figure 3. From the correlation coefficients obtained from the correlation module, it was exhibited that the parameters were separated by very small margins. In order to achieve superior feature extraction accuracy, a robust method is to be followed. The significant features for Kallooppara obtained are: P(t-1), Tmax(t-1), Tmax(t-2), Tmin(t-1), SM(t-3), SM(t-4). For Malakkara, P(t-2), P(t-3), Tmax(t-1), Tmax(t-2), Tmin(t-1), SM(t-1), SM(t-2). For Thumpamon: P(t-1), Tmax(t-1), Tmax(t-2), Tmin(t-1), SM(t-1).

The overall accuracy of the models was determined using the model performance evaluators. Highest R-squared value is obtained for Thumpamon station (0.981), and for Malakkara and Kallooppara, values obtained are 0.972 and 0.921 respectively. The R-squared value for a good prediction model will be nearby 1. As per this, it is concluded that the model is having good accuracy. The Root Mean Square Error (RMSE) obtained for Kalloopara, Malakkara, and Thumpamon stations were 0.291, 0.251, and 0.283 respectively. Based on the rule of thumb, the RMSE values between 0.2 and 0.5 exhibit that the model can predict the data accurately. The Mean Absolute Error (MAE) obtained was 0.271, 0.231, and 0.177 for the three stations respectively. The lower the MAE value, the better a model fits the dataset. The final evaluator, Mean Absolute Percent Error (MAPE), which is calculated by taking the mean of the absolute difference between actual and predicted values divided by the actuals, also manifests good accuracy of prediction for the model with values 10.02%, 9.80% and 6.20% for the three stations respectively. The Nash-Sutcliffe Efficiency (NSE) was also determined since it is the best criteria for the model performance evaluation aimed during low l and high streamflow periods. Higher NSE values were obtained for Hybrid model than that of the basic model for all the scenarios.

Based on the model evaluators acquired, it is clear that the hybrid model outperforms the basic LSTM model in prediction problems. Figure 4 depicts the variation of projected data by the basic

LSTM and hybrid RF-LSTM models with that of the actual data considering the test dataset for all three stations. This graph indicates that the streamflow prediction results of the Hybrid model fit the actual values better than the simple model. This also matches the statistical results reported before from model performance evaluators.

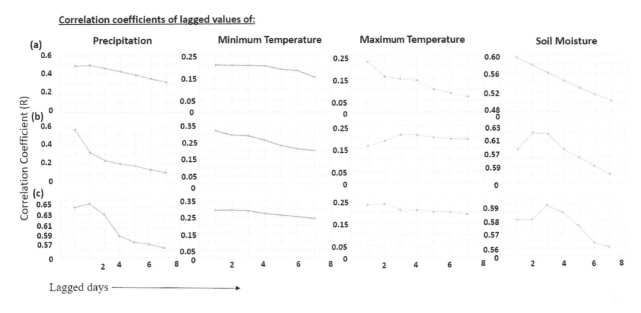

Figure 3. *Cross-correlation analysis results of (a) Malakkara (b) Thumpamon (c) Kallooppara.*

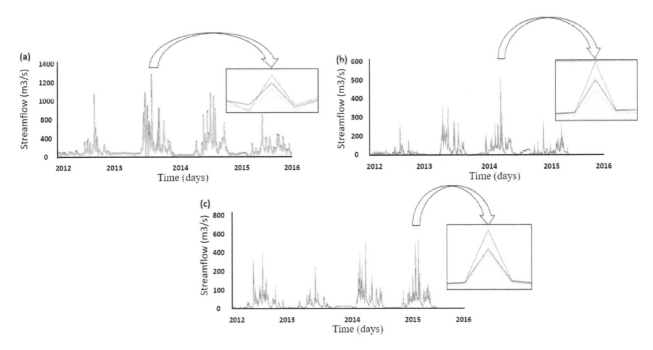

Figure 4. *Predicted and Actual Streamflowtest dataset (20% of whole data) of hybrid RF-LSTM model and Basic LSTM model for (a) Malakkara (b) Thumpamon (c) Kallooppara.*

This figure also depicts the satisfactory performance of the hybrid model over the standalone LSTM model in the extreme streamflow cases. For Malakkara, the highest value noted is 1213 m³/s, for which the hybrid model yielded a value of 989 m³/s and by basic model; it gives 810m³/s. For Thumpamon station, the highest extreme noted is 555m³/s and the hybrid and standalone LSTM models yielded 401m³/s and 306m³/s respectively. In the case of Kallooppara station, the highest extreme value is found to be 745m³/s and the yielded values from hybrid and standalone models are 512m³/s and 463m³/s respectively.

The results show that the presence of the LSTM's input-forget gate significantly improves the model's ability to predict non-linear time series data. The fine-tuning of model training methods, as well as the adoption of a trustworthy and long-term database, may improve the model's prediction accuracy.

4. CONCLUSIONS

Flood-related calamities can be avoided with proper streamflow forecasting. Using a hybrid RF-LSTM network, this study provided an efficient technique of streamflow prediction. To increase model performance and reduce model complexity, RF feature selection is used to discover the most important properties. It is contrasted to the basic LSTM model without any feature selection module to demonstrate its benefit in better time-series prediction with non-linear characteristics. When compared to a typical LSTM model that takes into account all inputs, the RF-LSTM model performs better in terms of many evaluation criteria such as R2, RMSE, MAE, and MAPE. Furthermore, the RF-LSTM model was found to represent the extreme high and low flows better than the standalone LSTM model, consistently across all stations. The future scope of the work deals with the incorporation of other climatic indicators for the prediction problems, such as governing climatic oscillations in the region, evapotranspiration and solar insolation, and so on, by determining the most relevant predictor set using the RF approach.

REFERENCES

[1] Hu, Y., Yan, L., Hang, T., and Feng, J. (2020). Stream-flow Forecasting of Small Rivers Based on LSTM.

[2] Ha, S., Liu, D., and Mu, L. (2021). Prediction of Yangtze River streamflow based on deep learning neural network with El Niño–Southern Oscillation. Sci Rep, 11(1), 11738.

[3] Choi, J., Won, J., Lee, O., and Kim, S. (2021). Usefulness of Global Root Zone Soil Moisture Product for Streamflow Prediction of Ungauged Basins. Remote Sens., 13(4), 756.

[4] Yilmaz, M. U., and Onoz, B. (2020). A Comparative Study of Statistical Methods for Daily Streamflow Estimation at Ungauged Basins in Turkey. Water, 12(2), 459.

[5] Mehdizadeh, S., and Sales, A. K. (2018). A Comparative Study of Autoregressive, Autoregressive Moving Average, Gene Expression Programming and Bayesian Networks for Estimating Monthly Streamflow. Water Resources Management, 32(9), 3001-3022.

[6] Chong, K. L., Lai, S. H., and El-Shafie, A. (2019). Wavelet Transform Based Method for River Stream Flow Time Series Frequency Analysis and Assessment in Tropical Environment. Water Resources Management, 33(6), 2015-2032.

[7] Sahay, R. R., and Srivastava, A. (2014). Predicting Monsoon Floods in Rivers Embedding Wavelet Transform, Genetic Algorithm and Neural Network. Water Resource Management, 28(1), 301-317.

[8] Adnan, R., Yuan, X., Kisi, O., and Yuan, Y. (2017). Streamflow Forecasting Using Artificial Neural Network and Support Vector Machine Models. American Scientific Research Journal for Engineering, Technology, and Sciences, 29, 286-294.

[9] Hassan, M., and Hassan, I. (2021). Improving Artificial Neural Network Based Streamflow Forecasting Models through Data Preprocessing. KSCE J Civil Engineering, 25, 3583–3595.

27. Effective Water Reservoir Management in Extreme Climatic Conditions for Flood Control - A Case Study on Kallada Irrigation Project

John K Satheesh,[1] Anand M,[2,*] and Suresh Babu D S[3]

[1,2] School of Environmental Studies, Cochin University of Science and Technology, Kochi, India

[3] National Centre for Earth Science Studies, Ministry of Earth Sciences, Thiruvananthapuram, India

*Corresponding authors:*anandm@cusat.ac.in, satheeshjohnk@cusat.ac.in, dssbabu@gmail.com

ABSTRACT: Kerala State with an enormous water resource system is more vulnerable to floods during extreme climatic events. 2018 cloudbursts followed by floods may recur in the current climate change scenario. In the wake of uncertain climate change combined with the drastic effects of long-term human interventions, the water resources sector in a particular region becomes an alarming turmoil. Against this backdrop, the left bank canal ayacut of Kallada Irrigation Project in Kerala was selected and the study was carried out by analyzing historical data of meteorology, and land use patterns. Data obtained from remote sensing and geographic information systems highlight changes in land use patterns and their effect on environment. The study puts forward suggestions for effective Reservoir management and sustainable utilization of water resources.

KEYWORDS: Water resource management, GIS and Remote Sensing, Climate change, Human intervention, Land use changes, Reservoir management

1. INTRODUCTION

In the wake of the unprecedented outbreak of climate change, various human interventions also impart vagueness to the uncertainties in the weather conditions. These climatic factors cause devastating and catastrophic events in the environment as a whole, especially in the water resources sector of a particular region. From the historical records of natural catastrophes, it is found that flooding is the most common natural hazard in the world and is reckoned as the third most damaging natural phenomenon globally after storms and earthquakes. From earlier times in history, water resources projects of various types were employed in different parts of the world to manage water resources.

Kerala, a south Indian state has a typical topography of lowland, midland, and highland and is covered with forests, vegetation, water bodies, and densely populated urbanized areas. Till recently, climate change was not a critical factor affecting the water resources sector. From the Indian Meteorological Department (IMD) data, it is noted that Kerala State in the southern part of India is getting an average precipitation of about 3000 mm in two monsoon spells. South West and North East monsoons are controlling the rainfall in the region, which occurs during a period of 6 months and accounts for about 90% of the annual rainfall. For the past 4–5 years, Kerala state has been experiencing untimely rainfall in an unexpected and distorted pattern. Kerala state is a land rich in water resources with 44 major and minor river basins and also includes 27 natural water bodies mostly coastal, numerous ponds, and 67 manmade reservoirs. While studying a recent a recent flood situation in Kerala, an extreme rainfall event (ERE) in the month of August 2018 which had a return period of 145 years, was identified to be the cause of a flood disaster in the region, particularly in the Periyar river basin in Kerala (Sudheer et al., 2019). Greenhouse gases and their impacts modify the precipitation and runoff pattern in a region, and cause sea level rise leading to land use and population shifts (Frederick and Major, 1997). Various climatic impacts reported by the Intergovernmental Panel on Climate Change (IPCC 1996) were analyzed and the vulnerability of water resource systems to climate variables was highlighted. Frederick and Major (1997) found that a meager change in any of these variables could lead to large changes in the whole system and its performance.

Land use is the way humans manage or alter land, while land cover is the spatial distribution of soil, water, vegetation, and anthropogenic activities on the Earth's surface. A region's land use and land cover (LULC) are the results of socioeconomic activities and natural changes in the surface of the earth. The hydrologic cycle is altered when land use features like forests, wetlands, and cultivated land are converted into urban areas. Such modifications slow down groundwater replenishment and accelerate surface runoff. Changes in land use and ecological shifts in land cover do not directly cause land degradation. Thoughtless land use practices, on the other hand, have a negative impact on climate and ecosystems and disrupt processes like biodiversity, water, and the release of trace gases (Riebsame et al., 1994). Understanding the landscape's dynamics and ensuring its sustainable management necessitates an in-depth analysis of LULC variation. Tools for geographic information systems (GIS) and remote sensing (RS) provide comprehensive insight into the spatiotemporal data of LULC. The scientific community has generally accepted these methods for precise discernment (Rawat et al., 2013, Misra and Balaji, 2015, Prasad and John, 2018). A compound data set of aerial photographs, satellite images, and maps can be used to create qualitative and quantitative geodatabases, as demonstrated by Kaliraj and Chandrasekar (2012). According to Wickware and Howarth (1981), researchers in this field discovered that RS methods have proven to be extremely useful in the timely and precise detection of regional and global deviations in LULC phenomena. According to Afify (2011), the Landsat multispectral scanner, Thematic Mapper (TM), and Enhanced Thematic Mapper Plus (ETM+) were utilized worldwide for the systematic detection of changes in land use. An effort has been made to investigate the efficacy of water resources projects by using tools of Geoinformatics in the Left Bank ayacut area of the Kallada Irrigation Project, Kerala state, India.

2. STUDY AREA

Kallada Irrigation Project (KIP) which was commissioned in the year 1986 is the largest irrigation project in the State of Kerala. The major water storage structure included in this project is the straight gravity masonry dam constructed in the downstream of the confluence of Kazhuthuruthy river, Senthuruny river, and Kulathuppuzha river, across the Kallada river, located at Parappar in Kollam District 80 57' N Latitude and 77 04' 20" E Longitude. The catchment area of Kallada dam is 549 Km² and the dam has a length of 335 m and a height of 85.35 m. The maximum water level (MWL) of the dam is 116.73 m and the full reservoir level (FRL) is 115.82 m with a gross storage capacity of 504.92 Mm³. Figure 1 depicts the study area map.

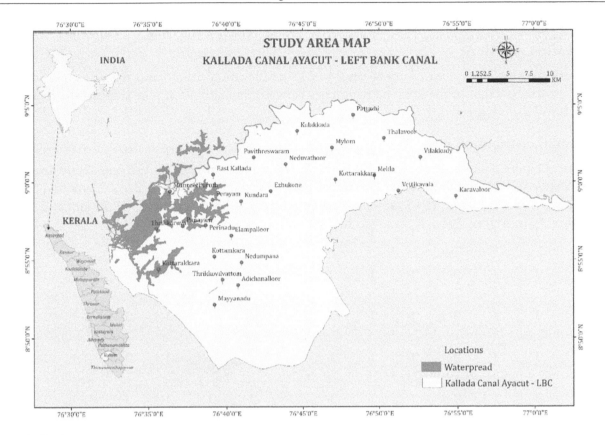

Figure 1. *Study area map.*

3. MATERIALS AND METHODS

For rainfall data, Climate Hazards Group Infrared Precipitation with Station data (CHIRPS) is used for deriving the rainfall distribution map for August 2008 and August 2018. The values from the CHIRPS data are extracted using Conversion Tool – Raster to point. Then Inverse Distance Weighted (IDW) spatial interpolation method is performed for mapping the spatial distribution of rainfall patterns [*Al-Mamoori et al., (2021)*]. ArcMap 10.8 version is the GIS Software used for mapping and analysis.

For assessing variations in LULC classes in the study area, geo-informatics and remote sensing (RS) are used. The online portals, Global Land Cover Facility (GLCF) and Earth Explorer, were used to obtain Landsat MSS, with a spatial resolution of 60 m, and Landsat 8 images, with a spatial resolution of 30 m, for the years 1990 and 2022 respectively. The land use maps were prepared using the RS software: ERDAS IMAGINE 9.2. The geometric modification of the imagery resamples the pixel grid to fit to a map projection to other reference image [*Kaliraj et al.,(2017)*]. In order to adapt the pixel grids and to remove the geometric alterations, Landsat images were registered and georeferenced to the Universal Transverse Mercator (UTM), World Geodetic System—WGS 84, 43 N coordinate systems, based on the topographic maps of 1:50,000 scale. This algorithm takes the pre-set value of the adjoining pixel and allocates it to the value of the output pixel, thereby relocating the original pixel values, without averaging, so that the subtleties and extremes of the pixel values are not lost (ERDAS Field Guide 1999).

Population data for the years 2000 and 2020 has been obtained from WorldPop. WorldPop is an interdisciplinary applied research group focusing primarily on supporting the improvement of the spatial demographic evidence base and the use of these data for health and development applications, including attaining the Sustainable Development Goals.

4. RESULTS AND DISCUSSION

4.1. Elevation and Slope

The elevation of the study area lies between 0 to 211 m. Elevation of midland region ranges from 30 m to 200 m whereas that of the lowland region lies between 0 m and 30 m. The slope of the region lies between 0 to 38.11 degrees.

4.2. Rainfall distribution map

Figure 2 and Figure 3 illustrate the spatial distribution of rainfall patterns for the years 2008 and 2018. During August 2008, the rainfall distribution for the region ranged between 221.02 mm and 345.40 mm (Figure 2), and in August 2018 rainfall distribution recorded was between 375.50 mm and 407.98 mm (Figure 3). In August 2018, Kerala was hit by devastating floods.

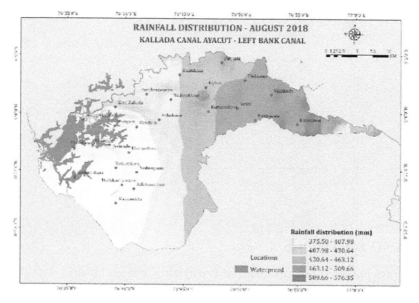

Figure 2. Rainfall distribution – August 2008.

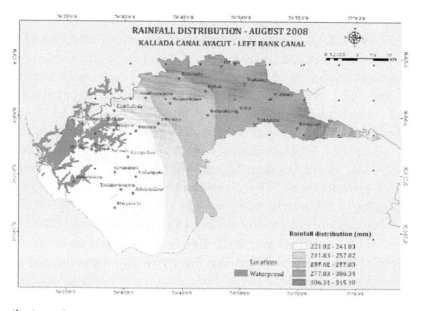

Figure 3. Rainfall distribution – August 2018.

4.3. LULC classification

For land use of 1990, Landsat 5 (a low earth orbit satellite launched on March 1, 1984) was used to collect imagery of the surface of the earth. Data from Landsat 5 was collected and distributed from the USGS's Centre for Earth Resources Observation and Science (EROS). The required imagery for the analysis has been downloaded from https://earthexplorer.usgs.gov/. For land use of 2022, Landsat 8 (launched on February 11, 2013, from Vandenberg Air Force Base, California, USA) was used. The imagery for the study has been downloaded from https://earthexplorer.usgs.gov/.

On analyzing the land use classes and area statistics, LULC maps derived for the years 1990 and 2022 are illustrated in Figures 4 and 5 respectively. Table 1 depicts the area of each land use class of 1990 and 2022 and also represented graphically in Figure 6.

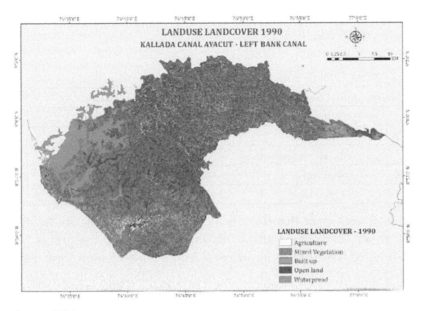

Figure 4. LanduseLandcover 1990.

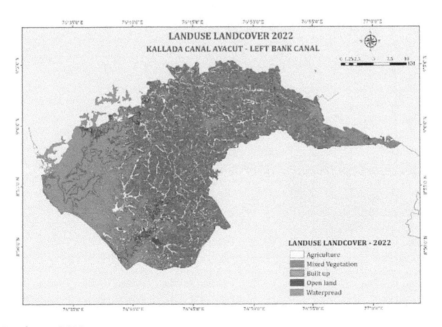

Figure 5. LanduseLandcover 2022.

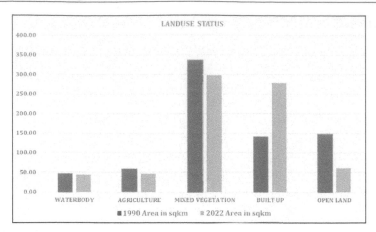

Figure 6. *Graph of LU classes for 1990 & 2022.*

Table 1. Land use areas (sq.km) for 1990 & 2022.

Sl. No.	LU class	1990	2022
1	Waterbody	47.11	43.84
2	Agriculture	59.03	46.43
3	Mixed vegetation	335.45	297.88
4	Built-up	140.13	278.24
5	Open land	146.09	61.41

From the analysis, it is very clear that there is a considerable decrease in the open land and mixed vegetation from 1990 to 2022. There is also a decrease in the land use status of water bodies and agriculture. Also, it is noticed that there is an increase in the built-up land form 140.13 sq.km in 1990 to 278.24 sq. km in 2022.

4.4. Population Density

Figures 7 and 8 depict the population density map for the years 2000 and 2020. From the map, it is evident that there is a considerable increase in the population density in both lowland and midland regions.

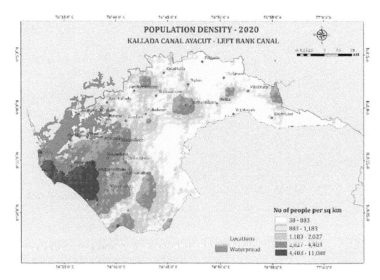

Figure 7. *Population density for the year 2000.*

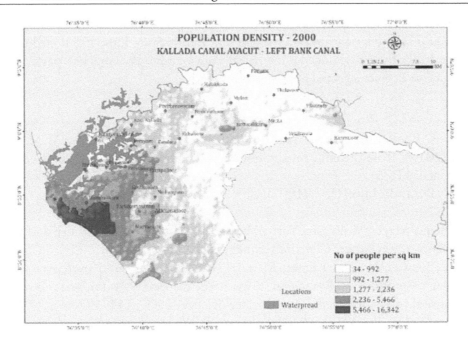

Figure 8. Population density for the year 2020.

5. CONCLUSION

While comparing land use data of 1990 and 2022, it was found that the area of open land, agriculture, mixed vegetation, and water bodies in the study area during 1990 got considerably reduced when it reached the year 2022. At the same time, there is an increase in the built-up area, mainly due to urbanization from 140.13 sq.km in 1990 to 278.24 sq.km in 2022. Population density in both lowland and midland regions in the study area had increased considerably during the period. The expansion of built-up land area is directly correlated with population growth. Remote sensing and geographic information systems highlight changes in land use patterns over time and how much they affect both the human population and the natural environment.

As considerable increase in rainfall distribution was observed in the year 2018, more uncertainties in weather conditions are likely to occur due to the phenomenon of unprecedented climate change. The extent of damages will be more drastic as the population density is rising in the area as noticed in the year 2020. In such cases, suitably located water resource projects with adequate storage capacity can play an effective role in storing the excess flown water and controlling the likely floods in the area. A well-thought-out plan for the use of land is necessary for harmonious ecological balance and sustainable development because the increase in the built-up area is directly correlated with the growth of the population.

In the case of the Kallada Irrigation project, the inferences brought out from this study can be used for planning and execution of effective water management in the area under investigation and also in regions with similar geomorphological characteristics. In a state like Kerala, where exists about sixty major water resources projects, several minor projects, and a large number of ponds, their effective maintenance and management can address sustainable development of the environment.

6. ACKNOWLEDGMENTS

The authors express thanks to the Director, NCESS, Thiruvananthapuram, India, and the Director, School of Environmental Studies, CUSAT, Kochi, India for supporting with all facilities for carrying out this study and the related reference work. Also, the authors thank the Irrigation Design and Research Board (IDRB) of Kerala State Irrigation Department and Kerala State Land use Board under the Kerala State Government, India for providing the historical data of rainfall, land use pattern, and reservoir operations of Irrigation Project selected for the project study.

REFERENCES

[1] Sudheer, K. P., Bhallamudi, S. M., Narasimhan, B., Thomas, J., Bindhu, V. M., Vema, V., and Kurian, C. (2019). Role of dams on the floods of August 2018 in Periyar River Basin, Kerala. Current Science, 116(5).

[2] Frederick, K. D., and Major, D. C. (1997). Climate change and water resources. Climatic Change, 37(1), 7–23.

[3] Riebsame, W. E., Meyer, W. B., and Turner, B. L. (1994). Modeling land use and cover as part of global environmental change. Climatic Change, 28(1), 45–64.

[4] Rawat, J. S., Biswas, V., and Kumar, M. (2013). Changes in land use/cover using geospatial techniques: A case study of Ramnagar town area, district Nainital, Uttarakhand, India. The Egyptian Journal of Remote Sensing and Space Science, 16(1), 111–117.

[5] Misra, A., and Balaji, R. (2015). Decadal changes in the land use/land cover and shoreline along the coastal districts of southern Gujarat, India. Environmental Monitoring and Assessment, 187(7), 1–13.

[6] Prasad, G., and John, S. E. (2018, April). Delineation of groundwater potential zones using GIS and remote sensing – A case study from midland region of Vamanapuram river basin, Kerala, India. AIP Conference Proceedings, 1952(1), 020028.

[7] Kaliraj, S., and Chandrasekar, N. (2012). Spectral recognition techniques and MLC of IRS P6 LISS III image for coastal landforms extraction along South West Coast of Tamilnadu, India. Bonfring International Journal of Advances in Image Processing, 2(3), 01–07.

[8] Wickware, G. M., and Howarth, P. J. (1981). Change detection in the Peace-Athabasca delta using digital Landsat data. Remote Sensing of Environment, 11, 9–25.

[9] Afify, H. A. (2011). Evaluation of change detection techniques for monitoring land-cover changes: A case study in new Burg El-Arab area. Alexandria Engineering Journal, 50(2), 187–195.

[10] Al-Mamoori, S. K., Al-Maliki, L. A., Al-Sulttani, A. H., El-Tawil, K., and Al-Ansari, N. (2021). Statistical analysis of the best GIS interpolation method for bearing capacity estimation in An-Najaf City, Iraq. Environmental Earth Sciences, 80(20), 1–14.

[11] Kaliraj, S., Chandrasekar, N., Ramachandran, K. K., Srinivas, Y., and Saravanan, S. (2017). Coastal land use and land cover change and transformations of Kanyakumari coast, India using remote sensing and GIS. The Egyptian Journal of Remote Sensing and Space Science, 20(2), 169–185.

[12] ERDAS Field Guide. (1999). Earth resources data analysis system. Atlanta, Georgia: ERDAS Inc. Retrieved from http://web.pdx.edu/emch/ip1/FieldGuide.pdf.

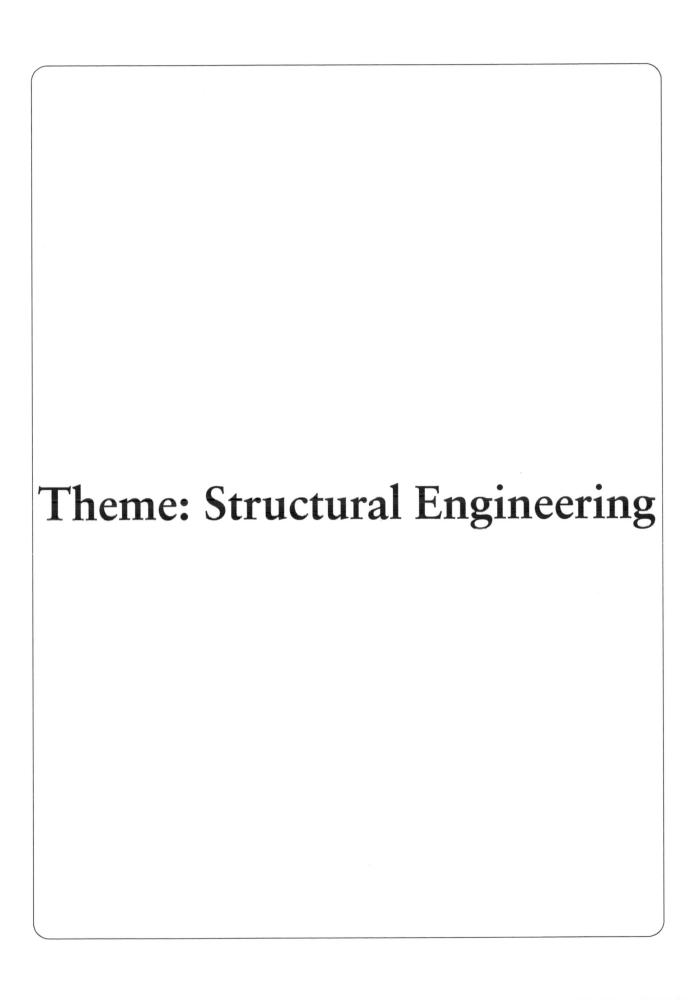

Theme: Structural Engineering

28. Seismic Behavior of High-Rise Building with and Without Shear Wall

Afnan Aslam[*] and Ajay Vikram

Department of Civil Engineering, Rayat Bahra University, Punjab, India

[*]Corresponding author: afnanaslam94@gmail.com

ABSTRACT: The major purpose of this article's research is to analyze seismic behavior and compare the findings of structures with and without reinforced concrete shear walls. Two structures with the same layout and an equal number of stories are evaluated, one with shear walls and one without. A brief overview of the design idea, as well as the requirement for a shear wall and the impact of an earthquake, is provided. On the same structures, response spectrum analysis was also undertaken. To account for the shear wall impact, the story displacements for various models are computed and compared. Model analysis and design are carried out using environment-friendly software ETABS 2019 in compliance with IS Codes.

KEYWORDS: Shear Wall, Stiffness, Storey Drift, Response Spectrum

1. INTRODUCTION

An earthquake is caused by a sudden release of energy in the Earth's crust, which causes seismic waves. The seismic activity of a region refers to the frequency, kind, and magnitude of earthquakes that occur throughout time. Vertical components that resist horizontal forces are known as shear walls. Shear walls are vertical components that carry and distribute earthquake and wind loads to the foundation. Shear wall systems are widely used to withstand lateral stresses caused by seismic excitation due to their high stiffness and strength. Shear barriers can effectively control drift against seismic loads.

2. MODEL CONFIGURATION

Two high-rise 21-story buildings with regular reinforced concrete structures are considered in seismic zone V and are located in Srinagar, India. The beam length in (x) transverse direction is 8 m, and beams in (y) direction are also of length 8 m. Figures 1 and 2 show the 3D and plan views of the 21-story buildings having 11 bays in x-direction and 9 bays in y-direction. Story height of each building is assumed 4 m. Beam cross-section is 450×850 mm and column cross-section is 1200×1600 mm.

The present study deals with seismic analysis using equivalent dynamic analysis of (G+20) story RC buildings using Structural Analysis and Design (ETABS) software.

Chapter 28 DOI- 10.1201/9781032657271-28

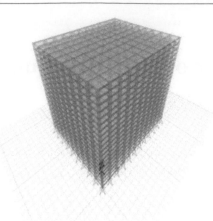

Figure 1. *3D view of 21-storey RCC building.*

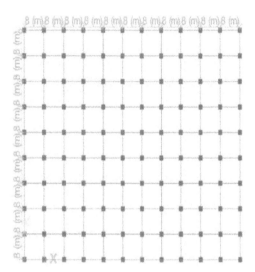

Figure 2. *Plan by determination of columns and beams.*

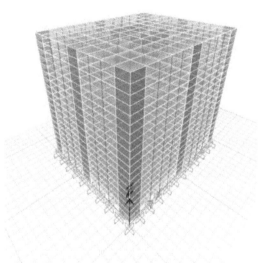

Figure 3. *3D view of building with shear wall.*

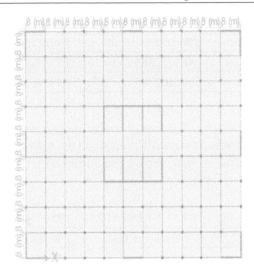

Figure 4. Plan of building with shear wall.

3. TIME PERIOD

Several S_a/g values are defined by IS-1893-2016 for various values of an approximate time period (T). For moment-resisting frame buildings without brick infill panels, the fundamental natural period (Ta) is calculated as Ta = 0.075h0.75 where h is the building's height in meters. (This is shown in the Table 1 below)

Table 1. Time period for building with shear wall and without shear wall.

Time period	With shear wall	Without shear wall
Global x	2.46 s	2.46 s
Global y	2.46 s	2.46 s

4. DESIGN BASE SHEAR

Using the code, IS-1893-2002 one may determine a building's design base shear:

$$V_b = A_h * W$$

where $A_h = \dfrac{Z I S_a}{2 R g}$

where W is the seismic weight and A_h is the design horizontal seismic coefficient.

The peak ground acceleration (z), importance factor (I), response reduction factor (R), and design acceleration coefficient (S_a/g) for various types of soil normalized equivalent to 5% damping determine the design horizontal seismic coefficient (A_h). The seismic zone V, medium soil, and 5% damping response spectrum analysis is performed using the spectra for medium soil as per IS 1893 (Part 1) 2002. These figures represent the spectral acceleration coefficients (S_a/g).

For medium soil sites,

S_a/g = 1 + 15T, (0.00 ≤ T ≤ 0.10), (T= time period in seconds)

= 2.50, (0.10 ≤ T ≤ 0.55)

= 1.36/T, (0.55 ≤ T ≤ 4.00)

S_a/g values for medium soil according to IS-1893-2002:

Table 2. Comparison of Design Base Shear of two different buildings.

Design base shear	With shear wall	Without shear wall
GlobalX (KN)	4647.1664	1949.8072
GlobalY (KN)	4860.1705	1894.1824

5. STIFFNESS

The force required to induce unit deformation/displacement in a structure is known as stiffness. This is quite evident in Table 3. and Table 4. that shows the values of stiffness in two buildings, with and without Shear Wall.

Table 3. Comparison of Story stiffness in x-direction of building with and without Shear Wall.

Story	With Shear Wall	Without Shear Wall
Story21	1227680	508573
Story20	1381498	1228143
Story19	1482575	1251401
Story18	1585220	1264875
Story17	1678840	1273926
Story16	1767401	1282296
Story15	1852962	1290421
Story14	1938149	1298612
Story13	2025946	1307026
Story12	2119832	1315901
Story11	2225127	1325328
Story10	2340574	1337849
Story9	2512603	1526922
Story8	2678568	1541771
Story7	2902680	1558958
Story6	3198674	1588130
Story5	3616874	1654464
Story4	4255396	1848429
Story3	5354901	2712356
Story2	7675323	9790680
Story1	16868653	9787738
Base	0	0

Table 4. Comparison of Story Stiffness in y-direction of building with and without Shear Wall.

Story	With ShearWall	WithoutShearWall
Story21	1227680	508338.4
Story20	1381498	1226435
Story19	1482575	1249737
Story18	1585220	1263255
Story17	1678840	1272349
Story16	1767401	1280750

Story15	1852962	1288895
Story14	1938149	1297098
Story13	2025946	1305513
Story12	2119832	1314382
Story11	2225127	1323796
Story10	2340574	1336297
Story9	2512603	1524914
Story8	2678568	1539730
Story7	2902680	1556866
Story6	3198674	1585953
Story5	3616874	1652073
Story4	4255396	1845557
Story3	5354901	2705164
Story2	7675323	9789972
Story1	16868653	9787049
Base	0	0

6. MAXIMUM STOREY DISPLACEMENTS

Storey displacement is the movement that takes place at each level of a storey. The top stories of multi-storey buildings will experience the greatest storey displacement. The storey displacement will attain its highest value as the height rises. (The comprison is in the Table 5 and Table 6 given below.)

Storey drift limitations = 0.004 h, (h= storey height) As per IS 1893:2016

Table 5. Comparison of Max displacement in y direction for building with and without Shear Wall.

Story	Without ShearWall	With ShearWall
Story21	53.177	36.322
Story20	47.833	33.869
Story19	45.45	31.54
Story18	42.961	29.24
Story17	40.366	26.98
Story16	37.672	24.754
Story15	34.891	22.563
Story14	32.037	20.409
Story13	29.122	18.295
Story12	26.159	16.229
Story11	23.159	14.218
Story10	20.133	12.273
Story9	17.096	10.402
Story8	14.406	8.641
Story7	11.72	6.976
Story6	9.045	5.43
Story5	6.407	4.021
Story4	3.864	2.771

Story3	1.578	1.707
Story2	1.23	0.86
Story1	0.196	0.269
Base	0	0

Table 6. Comparison of Max displacement in x-direction for buildings with and without Shear Wall.

Story	Without ShearWall	With ShearWall
Story21	53.09	34.999
Story20	47.737	31.158
Story19	45.358	29.013
Story18	42.872	26.888
Story17	40.281	24.788
Story16	37.589	22.713
Story15	34.812	20.665
Story14	31.961	18.65
Story13	29.05	16.675
Story12	26.09	14.748
Story11	23.094	12.878
Story10	20.071	11.075
Story9	17.038	9.355
Story8	14.352	7.735
Story7	11.669	6.215
Story6	8.998	4.813
Story5	6.364	3.544
Story4	3.828	2.428
Story3	1.552	1.487
Story2	1.12	0.745
Story1	0.187	0.232
Base	0	0

7. STOREY DRIFT

Storey drift is defined as the relative movement of any two floor levels of a storey between the floor above and below the floor under discussion. Horizontal displacement in a high-rise building is caused by drift. It affects movements of structural and nonstructural elements under seismic and wind load. Buildings wobble laterally when an earthquake occurs, and too much lateral displacement is undesirable. Large lateral displacements cause significant nonstructural damage, structural damage. Equations defining drift and drift index are,

Total drift of this floor = Δi Inter-storey drift of i floor (δ) $i = \Delta i - \Delta (i - 1)$

Drift index = deflection/height

Manual calculation storey drift limitations = $0.004h = 0.004 * 3*1000 = 12$ mm

As per the IS1893: 2016.

8. CONCLUSION

The building is designed using Indian standard (IS1893:2016) codes. We came at the following conclusions from our literature review:

- A high-rise building of 21 floors subjected to seismic, dead, and live loads were analyzed using ETAB 2019 software.
- Frame with shear wall at corners as in structure with shear wall performs better and the base shear gets increased when compared to the frame without shear wall.
- Shear wall performs better to lateral displacement and it reduces by more than 20% when compared to the frame without shear wall.
- The maximum story displacements in case of structure without shear wall are higher as compared to the structure with shear wall.

Story drift, also known as relative displacement, is the movement of one level with respect to the level below. It has been noted that the top drift decreases when a shear wall is present.

REFERENCES

[1] Banerjee, S. B., Barhate, P. D., and Jaiswal, V. P., "Aluform technology," vol. 2, no. 3, 2015, ISSN: 234-3696.

[2] Chandukar, P. P., and Pajgad, P. S., "Seismic analysis of RCC building with and without shear wall," *Int J Modern Eng Res*, vol. 3, no. 3, pp. 1805–1810, 2013, ISSN: 2249-6645.

[3] IS: 875 (Part 1), *Code of Practice for Design Loads (Other than Earthquake) for Building and Structures*, Bureau of Indian Standard, India, 1987.

[4] IS: 875 (Part 2), *Code of Practice for Design Loads (Other than Earthquake) for Building and Structures*, Bureau of Indian Standard, India, 1987.

[5] IS:13920:1993, *Indian Standard Ductile Detailing of Reinforced Concrete Structures Subjected to Seismic Forces–Code of Practice*, New Delhi: Bureau of Indian Standards.

[6] IS:456-2000, *Indian Standard Plain and Reinforced Concrete Code Of Practice*, Fourth Revision, New Delhi: Bureau of Indian Standards.

[7] IS:1893-2002, *Criteria For Earthquake Design of Structures*.

[8] IS:4326-1976, *Code Of Practice For Earthquake Resistant Design And Construction of Buildings*.

[9] SP: 22 *Explanatory Handbook on Codes For Earthquake Engineering*.

[10] Tidke, K., Patil, R., and Gandhe, G. R., "Seismic analysis of building with and without shear wall," *Int J Innov Res Sci Eng*.

29. Optimization of the Shapes of Buildings Using Aerodynamic Studies

Varada P N,[1] Neethu T P ,[1] Sankaranarayanan K M,[1,*] and Madhavan K T [2,*]

[1]Department of Civil Engineering, Sreepathy Institute of Management and Technology, Vavanoor, Palakkad

[2]R & D Centre, Sreepathy Institute of Management and Technology, Vavanoor, Palakkad, Kerala

*Corresponding authors: Sankaranarayanan. km@simat.ac.in,ktm@simat.ac.in

ABSTRACT: Aerodynamic characteristics of the forward and backward planes of the wake of a building can determine the intensity of the pollutants suspended in air, stability of the building etc. Hence it is beneficial to streamline the flow around the buildings. This could be achieved by altering the shape of the building optimally, preferably in the corners. Present study involved the aerodynamic characterization of differently shaped corners of a building. Four different configurations, viz., basic structure with setback, rectangular edges, chamfered, and circular corner smoothing were studied. Qualitative measurements (smoke flow visualization, tuft flow visualization) and quantitative measurements (pressure/velocity, drag) were carried out. Results of flow visualization studies indicated that the flow in the forward plane of the building is much more streamlined with modified corner shapes, in particular with corner smoothing. Building wakes can be effectively modified with chamfered and corner-smoothed body shapes. Corner-smoothed edge shows the most promising and beneficial results. Quantitative measurements (force measurements, wake survey) showed considerable improvement in the drag characteristics for corner-smoothed body shapes. It may also be noted that streamlining of the flow around the building could help to control the stagnation of suspended pollutants in front and behind the building, thus providing cleaner surroundings.

KEYWORDS: Building shape optimization, Flow around buildings, building models, building shape modification, wind tunnel measurements, flow visualization, wake survey, drag measurements

1. INTRODUCTION

Studies on modern tall structures and buildings have shown that they are sensitive to the wind forces acting on them. For the safety and endurance of buildings and structures, assessment of wind loads is essential before the design of buildings or structures. Aerodynamic considerations and architectural design have always attended to reduce the wind effect on buildings in parallel to reduce the vortexes at the pedestrian-level floor, corners, and behind the buildings. Architectural techniques are used to design aerodynamic tall buildings and comparisons are made among constructed prototypes using these techniques. Early study of aerodynamic impacts of buildings was done by Davenport (1971) using aerodynamic model tests. Trend towards tall buildings came in the 1990s, leading to research on corner correction, set-back and tapered aerodynamic forms, spoilers, and openings to obtain latent reduction in aerodynamic forces. Latent influences of aerodynamic corrections have been analyzed in economic terms (cost and available space) (Kwok, 1988; Dutton and Isyumov, 1990; Kareem et al., 1999; Cooper et al., 1997; Kim and Kanda, 2010; Isyumov et al., 1989). A leading study has been done by Daemei et al. (2017) and (2019) on designing a model of affordable housing in tall buildings with a focus on the use of environmental factors in humid subtropical climate-roofing.

Main objective of the present work was to make an assessment of the effectiveness of building shape optimization on the aerodynamic characteristics such as flow separation, wake width, wake defect, and drag on the forward and backward wake regions of buildings. These parameters can determine the intensity of the pollutants suspended in air, and the stability of the buildings. Data obtained from the study can be used effectively to determine the aerodynamic approaches for shape modification, including chamfered and rounded corners and thus to improve the building stability, hygienic factors and aesthetics.

Previous studies have shown that the flow field around an isolated building is characterized by complex three-dimensional structures, consisting of five main regions: incident turbulent flow, near-wake flow, turbulent wake, horse-vortex system and reattachment of the flow in the building faces. The interaction of the flow-building system depends on larger number of parameters, such as obstacle geometry, free-stream turbulence and approaching boundary layer flow. Optimization of the building shapes can be done by obtaining data of aerodynamic parameters like wake characteristics, drag forces, wind velocities etc. using qualitative methods and quantitative techniques. In the present work, flow visualization and pressure/velocity and drag measurements were done in the flow fields of four different configurations (rectangular, setback, chamfered and corner smoothed) building shapes. Results of the quantitative and qualitative measurements showed that the aerodynamic modification with the rounded corners and chamfered corners is capable of causing a reduction in the drag coefficient of the building. It was seen that building with rounded corners (with circular edge smoothing) is the optimal configuration and the most beneficial. It may also be noted that streamlining of the flow around the building can help to control the stagnation of suspended pollutants in front and behind the building, thus providing cleaner surroundings.

2. EXPERIMENTS

Models: Experiments were conducted in a neutral atmospheric boundary layer wind tunnel in the flow fields of the following building models shown in Figure 1: (a) rectangular, (b) corner-smoothed, (c) chamfered, and (d) setback.

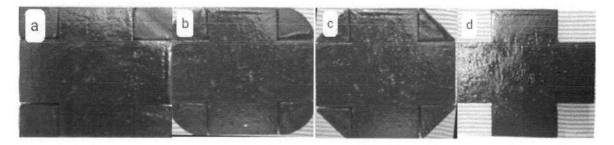

Figure 1. Model Configurations (Top View)

In order to standardize the measurements and normalize the measured data for the easiness of analysis, plan areas and heights of all configurations were kept to constant values of 1500 sqm and 50m respectively. An important aspect of all wind tunnel measurements is to maintain the geometric similarity between model and prototype, for applying similarity law. From considerations of wind tunnel test section geometry, a scaling factor of 1:500. To determine the effect of area on various aerodynamic parameters, four ratios length (l) to breadth (b) of 0.5, 1.0, 1.5, and 2.0 respectively were selected. Details are summarised in Table 1.

2.1. Wind Tunnels

Qualitative and quantitative measurements of aerodynamic parameters of the flow around scaled models were carried out in two wind tunnels: 0.15 m × 0.15 m Low Speed Demonstration Wind Tunnel at Spranktronics, Bangalore and 0.2 m × 0.2 m Low Speed, open circuit tunnel at CSIR-NAL, Bangalore.

2.1.1. *Quantitative measurements*

Quantitative measurements made include those of pressure in the wake regions behind and in front of the models under different conditions, aerodynamic forces on the building and subsequently the drag forces on different building shapes. Pressure measurements in the wake region were made using a Pitot-static probe placed at the point of interest in the flow.

Aerodynamic drag measurements: Strain gauge Balance Model CELRO M2 with measurement system consisting of (i) two bridges with variable gain, input from a four-arm Wheatstone bridge and a strain gauge transducer. Measured values of the Lift (zero in the present case) and Drag are directly digitally displayed.

2.1.2. *Qualitative Measurements*

Qualitative measurements provide most of the relevant information required to understand the effectiveness of modification of shape of buildings. Two techniques used for the implementation of the above are: (i) Tuft flow visualization where thin silk tufts of diameter of the order of 1-2mm are placed in the regions of interest in the flow field. Orientation and steadiness of the tufts are recorded to obtain qualitative information on the state of the flow (flow reversal, unsteadiness)

Smoke flow visualization where the illumination of the region of interest in the flow by a thin sheet of bright light source (flash lamp, high power LED, laser, etc.) along the axial direction of the flow. Illumination is done by a flash lamp through a thin slot, covered with glass, on the top window of the test section. A thin light sheet is obtained using light sheet optics, consisting of a combination of concave, convex, and cylindrical lenses. Flow visualization is done using fine smoke particles generated by vaporizing paraffin oil (slightly pressurized) dripping from an oil reservoir flowing over an electrically heated thin Nichrome wire, thereby obtaining a controlled stream of paraffin smoke that leaves streaks or traces in the stream, thereby visualizing the flow pattern.

3. RESULTS AND DISCUSSION

3.1. Tuft flow visualisation

As discussed earlier, tuft flow visualization was carried out in the wake of four configurations of the building model. Photographs of the visualizations of the four configurations are shown in Figures 2(a) to 2(d). As expected, the wake of the setback model is unsteady and is dominated by separated flow, as indicated by the reverse flow components (tufts along the middle portions going in backward direction). The unsteadiness and separation with flow reversals are introduced by the forward and backward-facing "steps" on the edges of the building model.

Figure 2. Tuft visualization in the wake of all configurations.

With rectangular shape Figure 2(b), the flow quality is seen to be improved. Region of reversed flow is smaller as indicated by the tufts; flow unsteadiness is also reduced, basically due to the reduction in the number of steps/sharp corners faced by the flow. In the case of models with chamfered corners, flow quality is seen to be improved further, although not dramatically, as shown in Figure 2(c). Reason for this could be the presence of sharp edges of the chamfered body. In the case of the corner-smoothed model, however, considerable improvement in the quality of the flow in the wake can be seen, as shown in Figure 2(d). The unsteadiness and flow reversals are greatly reduced and the wake can be seen much free of disturbances.

3.2. Smoke Wire Flow Visualisation

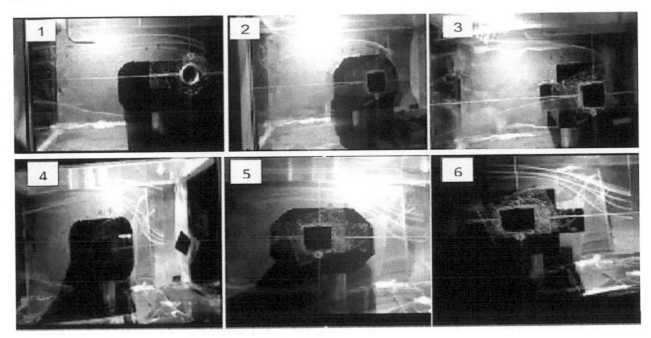

Figure 3. Smoke visualization in the wake of all configurations.

Photographs of the smoke flow visualization on three configurations, viz., rectangular, corner-smoothed, and setback respectively are shown in Figure 3 (Davenport, 1971; Kwok, 1988; Dutton and Isyumov, 1990) in the front region and Figure 3 (Kareem et al., 1999; Cooper et al., 1997; Kim and Kanda, 2010) in the wake. It can be seen that the flow is smooth in the case of the corner-smoothed model, both in the front and in the wake. Similar features can be seen in the flow visualization patterns around the model of L/b ratio 2.0 in the regions facing the flow, showing the flow being attached.

3.3. Pressure Measurements

Quantification of the results obtained from the flow visualization was done by determination of pressure distributions in the horizontal plane of the model wakes. From these measurements, velocities were calculated. Velocity profiles in the wake of the building models obtained thus are shown in Figure 4 variations in the normalized velocities, u/U_{max} *(where u is the local velocity and Umax is the freestream velocity),* in the horizontal (z) plane are plotted, as shown in the figure. It can be seen that wake-widths/wake-defects, indicative of aerodynamic forces of drag on the building models, depend largely on the shapes of the building models: Length and width of the wake being the least for the corner-smoothed model, followed by the chamfered and setback. This is in agreement with the results of flow visualizations, showing higher reductions in the unsteadiness and reversals in the flow caused by separation for chamfered and corner-smoothed models, in that order. An interesting observation is the shifts in the minimum values of mean flow velocity to the model center line (horizontal plane) with increasing uniformity. Clearly, for corner smoothed model, minimum peak is closer to the center, peak of setback model is the farthest, and in between for chamfered model. The flow non-uniformity in the free-stream flow region could have arisen from the errors in normalizing the local velocities (u) with freestream velocities (U_{max}), which had a certain amount of fluctuations. From the above observations, it can be further concluded that the corner-smoothed model causes the least amount of fluctuations and provides smaller and more uniform wake flow.

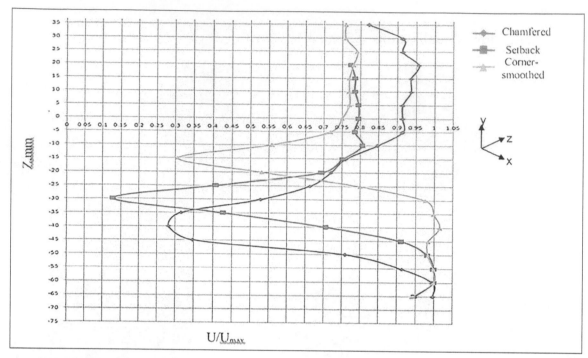

Figure 4. *Velocity profiles in the wake of building models*

3.4. Drag measurements

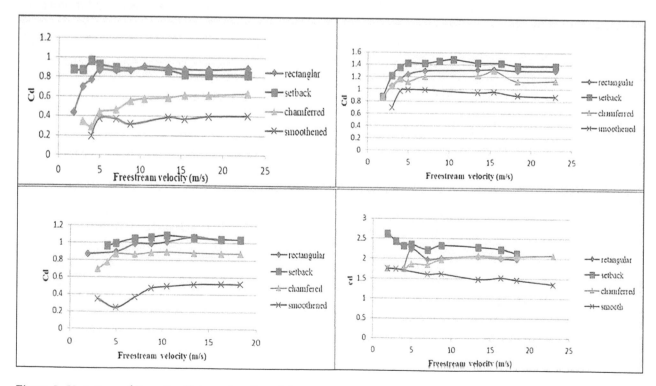

Figure 5. *Variations of Drag Coefficients for all models.*

As mentioned earlier, force/drag measurements on four model configurations with four l/b ratios were carried out for freestream velocities of 1m/s to 25m/s. Drags computed for each l/b ratio for all four models are shown in Figure 5. For comparison of the values of C_d for different cases of shapes, only the values when they reach a constant are to be considered. As expected, drag coefficient C_d is the least for corner-smoothed models and is reduced considerably for models with chamfered faces; graphs, drag increases with an increase in frontal area, irrespective of the l/b ratio of the models, percentage drag reduction is fairly similar.

4. CONCLUSIONS

Effectiveness of reducing adverse aerodynamic effects on buildings by optimizing the shapes of four building models was studied. Qualitative measurements (tuft flow visualization, smoke flow, visualization) and quantitative measurements (drag/force, pressure survey) were carried out on four different building shapes (rectangular, setback, chamfered, and corner-smoothed). While optimizing the shapes, main considerations were those of aerodynamic parameters of wake width and wake defect, drag, streamlining of flows, etc. Measurements showed that there is a considerable advantage in selecting a corner-smoothed model for the stability of building, reduced size of building wakes, and better distribution of suspended particles, including pollutants by convection. As drag increases with increased frontal area, it may be a good idea to design buildings with a nominal frontal area. Thus, a building with lower frontal area (lower l/b ratio) is preferable from the considerations of reduced disturbances, smooth airflow around the building. Reduced wake width and flow uniformity achieved thus can bring in reduction of stagnant pollutant air particles, bringing in more environmental-friendliness. This has high relevance and applicability in the cases of high-rise buildings where pollutant-filled building wakes are causes of concern.

5. ACKNOWLEDGMENTS

The authors are grateful to the Principal, SIMAT, and the Head, Department of Civil Engineering, SIMAT for their guidance and encouragement during the work. They are also very thankful to Dr. G. Ramesh, Head, Aeronautical Engineering Department, Gopalan College of Engineering and Management, Bangalore for the help and guidance throughout the work. The support given to us by Sri. M. G. S. Vithal and Mr. M. V. Nagaraj, Spranktronics, Bangalore is thankfully appreciated. The support and guidance given to us by Dr.P.V.S Murthy, Head, UAVID Division, CSIR-NAL, Bangalore, and the staff of Microair Vehicle Aerodynamic Research Tunnel for permitting us to carry out the work there is gratefully acknowledged.

REFERENCES

[1] Davenport, A. G. (1971). The response of six building shapes to turbulent wind. Philosophical Transactions of the Royal Society of London. Series A, Mathematical and Physical Sciences, 269(1199), 385–394.

[2] Kwok, K. C. S. (1988). Effect of building shape on wind-induced response of tall building. Journal of Wind Engineering and Industrial Aerodynamics, 28(1–3), 381–390.

[3] Dutton, R., and Isyumov, N. (1990). Reduction of tall building motion by aerodynamic treatments. Journal of Wind Engineering and Industrial Aerodynamics, 36, 739–747.

[4] Kareem, A., Kijewski, T., and Tamura, Y. (1999). Mitigation of motions of tall buildings with specific examples of recent applications. Wind and Structures, 2(3), 201–251.

[5] Cooper, K. R., Nakayama, M., Sasaki, Y., Fediw, A. A., Resende-Ide, S., and Zan, S. J. (1997). Unsteady aerodynamic force measurements on a super-tall building with a tapered cross section. Journal of Wind Engineering and Industrial Aerodynamics, 72, 199–212.

[6] Kim, Y., and Kanda, J. (2010). Characteristics of aerodynamic forces and pressures on square plan buildings with height variations. Journal of Wind Engineering and Industrial Aerodynamics, 98(8–9), 449–465.

[7] Isyumov, N., Dutton, R., and Davenport, A. G. (1989). Aerodynamic methods for mitigating wind-induced building motions. In Structural Design, Analysis and Testing, ASCE, pp. 462–470.

[8] Daemei, A. B. (2017). Thesis: Designing a model of affordable housing in high-rise building with a focus on the use of environmental factors in humid subtropical climate - Roofing Rasht.

[9] Daemei, A. B., Khotbehsara, E. M., Nobarani, E. M., and Bahrami, P. (2019). Study on wind aerodynamic and flow characteristics of triangular-shaped tall buildings and CFD simulation in order to assess drag coefficient. Ain Shams Engineering Journal, 10(3), 541–548.

30. Comparative Study of Triangular Hollow Flange I Section and Conventional I Section Under Shear

Vismaya Ravindran and Divya B Mathew
Department of Civil Engineering, Government Engineering College, Thrissur, India
Corresponding author: vismayaravindranc@gmail.com, divyabmathew@gectcr.ac.in

ABSTRACT Cold-formed stainless-steel sections are often used in building construction because of their superior corrosion resistance, ease of maintenance, and pleasing appearance. The corrosion resistance of cold-formed stainless steel makes stainless-steel structural components more durable. Being a recyclable material points out stainless steel as a sustainable solution to construction wastes. A cold-formed thin-walled steel hollow flange beam (HFB) has been developed and can be utilized structurally. It is composed of one or two closed flanges with high torsional stiffness and a relatively flexible web. In this article, a comparative study of triangular hollow flange I section, rectangular hollow flange I section, and conventional I section is presented. Numerical model developed using Ansys software is validated against literature available. The triangular and rectangular hollow flange section and conventional I sections are all modeled with and without web side plates, three web side plates are provided at each side. In the absence of a web side plate, buckling of web element is present in rectangular hollow flange and conventional I section. In general, the section having web side plates has a higher load-carrying and shear capacity compared to the other sections without a web side plate. The load carrying capacity was higher in the case of the triangular hollow flange section when compared with rectangular hollow flanges and conventional I section with or without the presence of a web side plate. The global stability of triangular hollow flange beams has greatly improved because of their superior torsional stiffness and stability compared to conventional I beams.

KEYWORDS Cold-Formed Stainless-Steel Section, Hollow Flange Beam, Triangular Hollow Flange Beam, Rectangular Hollow Flange Beam

1. INTRODUCTION

Cold-formed stainless-steel sections are often used in building construction due to their higher corrosion resistance, ease of maintenance, and pleasing appearance. The main feature of cold-formed stainless steel is its corrosion resistance which makes stainless-steel elements more durable. Because it is a recyclable material, stainless steel stands out as a sustainable solution to construction waste. Even though stainless steel is approximately four times more expensive than standard normal carbon steel, it was proposed that stainless-steel structures are more economical on the basis of whole life than carbon steel in severe situations. The stress–strain curves for tension and compression can also differ significantly. This has implications on the buckling behavior of members and the deflection of beams. Cold-formed stainless-steel sections are more frequent in light structural applications than hot-rolled and built-up sections. Cold-formed thin-walled steel hollow flange beam (HFB) is made up of one or two torsional rigid closed flanges and a reasonably flexible web. The cross-sections of doubly symmetric rectangular hollow flange beams (RHFBs), triangular hollow flange beams (THFBs), and lite steel beam (LSB) also known as hollow flange channel beam as shown in Figure 1. Electric resistance welding can be used to join the cold-formed hollow flanges to the web parts to create the sections. Closed flanges limit distortional buckling effects, and doubly symmetric hollow flange I sections were more stable to torsional impacts than monosymmetric light steel beam sections. Hollow flange beams combine the stability of cold-formed steel sections with a high strength-to-weight ratio, making them preferable to conventional sections. The hollow flanges section is away from the center, making them more efficient flexural members than equivalent cold-formed sections.

Figure 1. *(a) Rectangular hollow flange I section, (b) triangular hollow flange I section, (c) rectangular hollow flange channel section.*

2. FINITE ELEMENT MODELLING

A finite element package, Ansys Workbench 2019 R19.0 is used to simulate the hollow flange I section. For validation experimental study conducted by Poologanathan and Mahendran (2010) was chosen. The article conducted a detailed experimental study of shear response of cold-formed stainless-steel sections in a Lite steel beam section.

2.1. Development of FE model

For validation, the RHFB section having aspect ratio 1, subjected to point load is considered. Geometrical properties of the section are given in Table 1. In the article, geometrical imperfection is taken as L/150. Aspect ratio is the ratio of shear span (a) to the depth of a web element (d). Three web side plates (WSPs) are provided at the supports and loading point on each side. The entire rectangular hollow flange section has a 5 mm mesh, and all six web side plates have a 10 mm mesh.

Table 1. Geometrical details of RHFB section

Specimen details	Dimensions (mm)
Overall depth	200
Depth of the flange	20
Width of the flange	60
Thickness	2
Overall length	495
Shear span	160
Width of web side plate	75

2.2. Material Properties

Cold-formed stainless-steel sections of density 7750 kg/m3, Young's modulus of 200,000 MPa, and a Poisson's ratio of 0.3 is chosen. Average yield stress and ultimate tensile strength of top flange, bottom flange, and web were 479.5 MPa and 537.5 MPa, 541.9 MPa and 595.1 MPa, and 446.4 MPa and

531.1 MPa, respectively. Bonded connections were used to connect the web side plate with the web element. To ensure simply supported circumstances, pin support boundary conditions were used on both sides of the beam ends. To avoid torsional effects, the in-plane translational degrees of freedom of the cross-sectional plane (X-Y-Z plane) were restrained for the application of pin-supported conditions to the beam sections, and the rotational degrees of freedom about the longitudinal axis (Z-axis) of the section was restrained. The other axes, X and Y are unrestricted.

2.3. Model Validation

The ultimate load and deformation values from experimental and numerical are given in Table 2. The percentage variations of ultimate load and deformations were 3.27% and 7.18%. The load-versus displacement responses of the numerical models of the loading condition are presented in Figure 2.

Table 2. Validation

Parameter	Experimental values (Poologanathan and Mahendran, 2010)	FE model	Percentage variation (%)
Load	295.2 kN	285.52 kN	3.27
Deformation	5.81 mm	6.26 mm	7.18

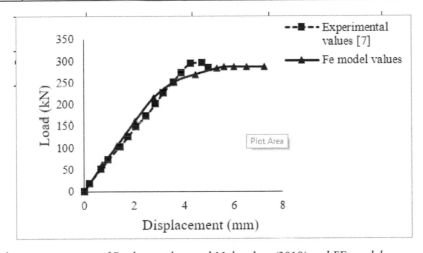

Figure 2. *Load displacement response of Poologanathan and Mahendran (2010) and FE model.*

Failure mode of experimental specimen and numerical model are shown in Figure 3(a) and (b), respectively. It can be concluded that the developed numerical model very well predicted the force–displacement relationship and also the failure mode of the rectangular hollow flange sections with web side plates.

3. NUMERICAL STUDY

3.1. General

The effect of the web side plate on ultimate load, shear response, and failure pattern of rectangular hollow flange I section, triangular hollow flange I section, and conventional I section was studied. Slender cold-formed stainless-steel sections of aspect ratio 1 and austenitic grade 1.4301. The material properties were Young's modulus of 196.1 GPa, yield stress of 298 MPa, ultimate stress of 632.5 MPa, and ultimate strain of 0.4. Section 450 × 90 × 40 × 3.8 mm was selected and details of the section were given in Table 3. The three web side plates were placed at each side of the sections and the distance between two web side plates was 370 mm. The dimensions of the web side plate were 75 mm in width and 3 mm in thickness.

Figure 3. *Failure pattern of rectangular hollow flange section (a) test specimen (Poologanathan and Mahendran, 2010) and (b) FE model.*

Table 3. Geometrical details of THFB, RHFB, and I section.

Section	Span (mm)	Overall depth (mm)	Depth of flange (mm)	Width of flange (mm)	Thickness (mm)	Shear span (mm) (a)	Depth of web (mm) (d)	Aspect ratio ($^a/_d$)
THFB	915	450	40	90	3.8	370	370	1
RHFB	915	450	40	90	3.8	370	370	1
I section	915	450	–	90	3.8	370	370	1

4. RESULT AND DISCUSSIONS

4.1. Behavior of Rectangular and Triangular Hollow Flange Section and Normal Section Without Web Side

PlateTriangular hollow flange and rectangular hollow flange I sections as well as the I sections are designed without web side plates. In the absence of web side plates, buckling in web as well as local buckling of flange and web was present in the case of conventional I section and rectangular hollow flange I section. In the case of triangular hollow flange section, it can be seen that only local buckling of flange is present. Triangular hollow flange I sections have shown the ability to resist web buckling, Different buckling failure of the three sections are shown in Figures 4–6. Web buckling occurs when the intensity of vertical compressive stress near the center of the section becomes greater than the critical buckling stress for the web acting as a column. A heavy concentrated load or end reaction produces a portion of high compressive stresses in the web either at support or under the load. This makes the web either to buckle or to cripple. Flange buckling which is a common failure pattern in triangular hollow flange I sections, rectangular hollow flange I sections, and conventional I sections, occurs when the width-to-thickness ratio of flange is insufficient to withstand the moment on the beam. Limiting the width-to-thickness ratio was one approach to preventing this form of buckling. Initially, the web of the section takes the shear load, the flange takes the bending moment. The shear capacity of the web is independent of the bending moment actions. When the bending moment is greater than the bending capacity of the flange section, the yielding of the flange starts and buckling of the flange occurs. plotted for triangular hollow flange and rectangular hollow flange I sections as well as conventional I sections, and it can be seen that triangular hollow flange I sections carry more load as compared to rectangular hollow flanges I sections and conventional I sections. The ultimate loads of triangular and rectangular hollow flanges I sections and normal I sections are 134.23, 123.34, and 102.68 kN, respectively.

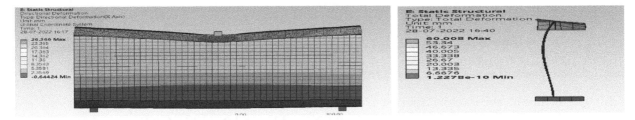

Figure 4. *Failure pattern of conventional I section.*

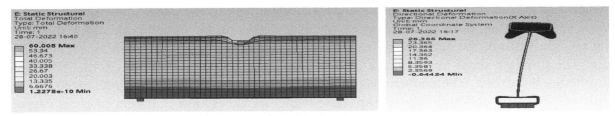

Figure 5. *Failure pattern of Rectangular hollow flange section.*

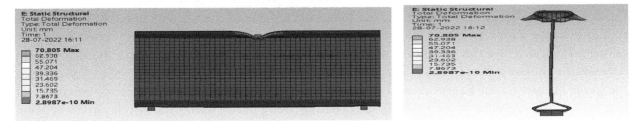

Figure 6. *Failure pattern of triangular hollow flange I section.*

The ultimate shear capacity of triangular and rectangular hollow flanges I sections as well as conventional I section when calculated are 67.115, 61.67, and 51.34 kN, respectively. So, in the absence of web side plate, the ultimate shear capacity of the triangular hollow flange section was 16.75% and 23.5 % higher than the rectangular hollow flange I section and conventional I section.

Behavior of Rectangular and Triangular Hollow Flange I Section and Normal I Section With Web Side PlateSix web side plates were provided for each of the three sections. The distance between the two-web side plate is taken as 320 mm. The failure patterns of the above three sections are different as shown in Figures 7–9, respectively. A diagonal tension field is present in the case of the rectangular hollow flange section. Due to relatively higher torsional stiffness, diagonal tension fields are absent in the case of triangular hollow flange I section. In the case of a conventional I section with web side plate, shear buckling is present on both sides of the center web side plate, but lateral displacement of the web is controlled. In rectangular hollow flanges I section, shear buckling is present on one side of the web side plate and the flange has no deformation. But in the triangular hollow flange I section, deformation is only present at the loading point. The ultimate loads of triangular and rectangular hollow flanges and normal I sections are 243.09, 223.36, and 201.59 kN, respectively. The ultimate load capacity of the triangular hollow flange I section is 8% and 17.07 % higher than the rectangular hollow flange I section and conventional I section. Ultimate load-carrying capacity of section greatly improved in the presence of web side plate and web buckling was also controlled.

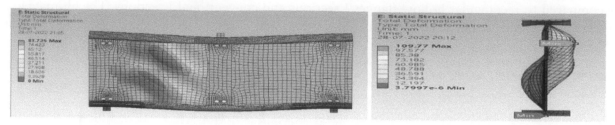

Figure 7. Failure pattern of conventional I section with web side plate.

But the ultimate shear capacities of triangular and rectangular hollow flange I sections and conventional I sections are 121.545 kN (243.09 kN/2), 111.68 kN (223.36 kN/2), and 100.795 kN (201.59 kN/2), respectively. The ultimate shear capacity of triangular hollow flange I section is 8.11 % higher than that of rectangular hollow flange sections and 17.07 % higher than normal I section. Inelastic shear buckling occurs in all cases of different shapes of the section with and without web side plate. The ratio of the depth of the web to the thickness of the web governs the type of shear failure. The ultimate shear and ultimate load of triangular hollow flange I sections, rectangular hollow flange I sections, and conventional I sections are increased due to the presence of a web side plate. Triangular hollow flange sections have higher shear response and load-carrying capacity compared to other sections, such as rectangular hollow flange sections and conventional I sections. Local buckling of flange is seen as the failure mode of triangular hollow flange section with and without web side plate. Local buckling of the flange occurs due to the bending moment being higher than the bending capacity of the flange sections. In triangular hollow flange sections, increased anchoring facilitated by the triangular flanges and transverse stiffeners that cause the distribution of stresses in the web more evenly. Web buckling can be seen as the failure mode of present in rectangular hollow flange I sections and conventional I sections in the presence and absence of web side plates. In the presence of web side plates, shear buckling of the web occurs in the space between two web side plates.

Figure 8. Failure pattern of triangular hollow flange I section with web side plate.

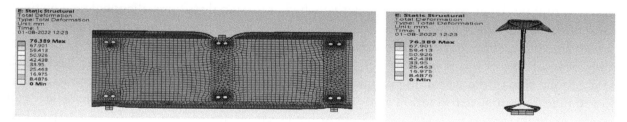

Figure 9. Failure pattern of rectangular hollow flange I section with web side plate.

Table 4. Comparison of RHFB, THFB, and I section with and without web side plate.

Types of section	Ultimate load (kN)		Ultimate shear (kN)	
	With WSPs	Without WSPs	With WSPs	Without WSPs
Conventional I section	201.59	102.68	100.795	51.34
RHFB	223.36	123.34	111.68	61.67
THFB	243.09	134.23	121.45	67.115

5. CONCLUSION

A numerical investigation was conducted on triangular, rectangular hollow flange as well as conventional I section. The model was created in Ansys software and validated using experimental data from Poologanathan and Mahendran (2010).

- Load carrying capacity is more for triangular hollow flange section as compared to rectangular hollow flange and conventional I section.
- The triangular, rectangular hollow flange I section and conventional I sections are modeled with and without web side plates and it was observed that on providing web side plate the ultimate load carrying capacity was increased by 49.06%, 44.77%, and 45.68%, respectively when compared to the section without web side plate.
- Web buckling of conventional I-section was controlled by the presence of a web side plate, so the web side plate had the ability to resist buckling.
- The sections have a moderate ratio of depth of web to thickness of web, so the buckling is inelastic shear buckling as per AS/NZS 4600.

REFERENCES

[1] Dissanayake, D.M.M.P., Zhou, C., Poologanathan, K., Gunalan, S., Tsavdaridis, K. D., and Guss, J. (2021) "Numerical simulation and design of stainless-steel hollow flange beams under shear," Journal of Constructional Steel Research, Vol. 176, pp. 399–414.

[2] Eduardo Carlos and Gonqalves Carvalho (1991) "The behaviour of cold-formed stainless steel beam webs subjected to shear and the interaction between shear and bending," University of Johannesburg.

[3] Gardner and Theofanous, M. (2008) "Discrete and continuous treatment of local bucking in stainless steel elements," Journal for Construction of Steel Research, Vol. 64, pp. 1207–1216.

[4] Poologanathan Keerthan (2014) "Shear behaviour and design of lite steel beam," Journal of Constructional Steel Research, Vol. 25, pp. 256–262.

[5] Poologanathan Keerthan and Mahen Mahendran (2010) "Experimental studies on the shear behaviour and strength of LiteSteel beams," Engineering Structures, Vol. 32, pp. 3235–3247.

[6] Poologanathan Keerthan, Mahen Mahendran, and David Hughes (2014) "Numerical studies and design of hollow flange channel beams subject to combined bending and shear actions," Engineering Structures, Vol. 77, pp. 129–140.

[7] Wanniarachchia, K. S. and Mahendran, M. (2017) "Experimental study of the section moment capacity of cold-formed and screw-fastened rectangular hollow flange beams," Thin-Walled Structures, Vol. 119, pp. 499–509.

31. Numerical Investigation on the Fatigue Strength of Short CFDST Column

Farzana O A[1] and Miji Cherian R[2]

[1]M.Tech. Scholar, Department of Civil Engineering, Government Engineering College Thrissur, Kerala, India

[2]Associate Professor, Department of Civil Engineering, Government Engineering College Thrissur, Kerala, India

*Corresponding author: farzanaayoob60@gmail.com, mijicpaul@gectcr.ac.in

ABSTRACT: Concrete filled double steel tubes (CFDST) columns are composite columns composed of concrete sandwiched between outer and inner steel tubes leaving a central hollow portion. The study aims to investigate numerically the fatigue behavior of circular CFDST short columns under reversed cyclic loading using a validated finite element model. A parametric study is undertaken, investigating the fatigue performance of a series of circular CFDST columns under the effect of parameters like axial load, drift, and thickness of the outer steel tube. The critical drift and optimum axial loading condition for efficient performance and the trend in reduction of fatigue life due to loss of thickness of outer steel tube are also determined.

KEYWORDS: CFDST, Fatigue analysis, Finite element analysis, Numerical model

1. INTRODUCTION

CFDST columns are composed of concrete sandwiched between outer and inner steel tubes leaving behind a central hollow portion. In CFDST, the confinement of concrete is provided by the steel tube and the local buckling of the steel tube is improved due to the support of the concrete core. Uenaka et al. (2009) experimentally investigated the mechanical behavior of CFDST stub columns under axial compression and evaluated the confinement effect by the outer tube to the filled concrete strength and proposed equations to estimate their ultimate strengths under compression based upon the yield strengths of the tubes and the filled concrete cylinder strength. Ayough et al. (2019) reviewed the models in the existing literature for predicting the behavior of steel materials and confined concrete. They concluded that the European and Australian codes gave the most reliable predictions, while American design codes gave conservative predictions on the ultimate strength of CFDST column. Dehghani and Aslani (2019) provided a comprehensive review on common damages or deterioration of fixed steel jackets used as substructures in offshore structures. They concluded that fatigue failure (around 24.9 %) is one of the major concerns for offshore platforms as they are extensively subjected to repeated forces from waves and wind during their service life. The majority of the studies have been conducted to propose the equations for the estimation of the strength of CFDST columns. But, the studies on the evaluation of fatigue strength and corrosion of CFDST are sparse. Therefore, to be effectively introduced into the construction industry, the fatigue strength and corrosion potential of CFDST need to be investigated. So, in the current investigation, the study on the behavior of short CFDST columns under reversed cyclic loading is conducted using the finite element (FE) package ANSYS Workbench.

This paper presents a numerical study conducted to investigate the behavior of short CFDST columns under reversed cyclic loading. The details of the study on the fatigue behavior of circular CFDST short column under reversed cyclic loading using validated finite element model and the parametric study on the fatigue performance of a series of circular CFDST columns under the effect of parameters like axial load, drift, and thickness of outer steel tube are presented herein. The critical drift and optimum axial loading condition for efficient performance and the trend in reduction of fatigue life due to loss of thickness of outer steel tube are also presented.

2. NUMERICAL ANALYSIS

2.1. Experimental data for validation of numerical model

The properties, dimensions and test setup of the experimental model by Dehghani and Aslani (2019) is used for numerical modeling. The height of the specimen is 1m. The Poisson's ratio of steel and concrete is 0.3 and 0.18 respectively. The stress–strain curve of concrete is taken from Park and Paulay (1975) and the typical stress–strain curve for steel's behavior is taken from by Pagoulatou et al. (2014).

2.2 Validation of numerical model

For the simulation of circular concrete-filled double-skin steel tubes (CFDST) and the analysis of their behavior, the finite element analysis was used. The finite element analysis of the model was done and the results were compared with the experimental results to validate the numerical model. In the finite element analysis, a general type of eight-node solid element (solid 186) is used for both the outer steel tube and inner steel tube and sandwiched concrete. The multilinear isotropic strain hardening is used to provide the tabular stress-strain data of concrete and Bilinear isotropic strain hardening for steel. A bonded contact was provided between outer steel tube and inner steel tube with top and bottom plate. Frictional contact with friction values of 0.3 and 0.4 was provided between concrete core and steel tubes. All nodes on the bottom surface of the column were assumed to fixed, whereas all nodes on the top surface of the column were set free only in the longitudinal direction. A displacement along the longitudinal direction was applied on the top plate in the downward direction.

2.2.1 Verification of finite element model

The different parameters used for validation are ultimate axial capacity vs. strain curve, directional deformation, failure pattern, and ultimate capacity. The axial capacity vs. strain graph of numerical models with friction coefficient of 0.3, 0.4 and the experimental analysis by [4] are shown in Figure 1. CFDST numerical model with friction coefficient of 0.4 showed a more proximity to the experimental results. The ultimate capacity of CFDST column obtained is tabulated in Table 4. In the experimental analysis, the directional deformation of CFDST column was obtained as 4mm by the yielding of outer tube in an outward direction in the topmost part of column. The directional deformation obtained from numerical model is 3.56mm in the topmost region of column which is shown in Figure 8. The failure pattern of the experimental model of CFDST column is by buckling of outer tube outwards in upmost region of column. The numerical model also showed a similar failure pattern which is as shown in Figure 2. Taking into account the aforementioned points, it can be concluded that the developed numerical model is reliable and can accurately capture the real behavior of the CFDST members.

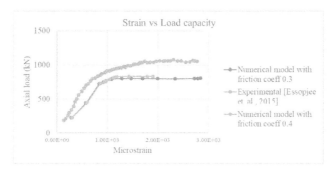

Figure 1. Axial load vs. micro strain graph.

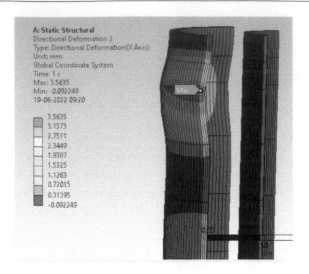

Figure 2. *Directional deformation and failure pattern of CFDST numerical model.*

2.3 Numerical modelling of fatigue

The fatigue analysis decisions taken are given in Table 1. The significant results obtained from numerical model of CFDST short column loaded to 650kN axial load (70% ultimate capacity) and 0.5 % drift is described below. Percentage drift refers to the percentage of lateral displacement to height of specimen.

Table 1. Fatigue analysis decision.

Sl. No.	Parameters	Decision
1	Analysis type	Stress life
2	Loading	Constant amplitude proportional loading
3	Mean stress correction	None
4	Strength factor	1
5	Multiaxial stress correction	Von Misses criteria

2.3.1 Fatigue capacity of CFDST column

2.3.1.1. Fatigue life

If loading is of constant amplitude, fatigue life represents the number of cycles until the part fails due to fatigue. The fatigue life of outer steel, inner steel, and concrete core are tabulated in Table 2. From the analysis results, it is obvious that the failure due to fatigue occurs at first on the concrete core due to combined bending and compressive force. But the crushed concrete is contained in the double confinement of outer and inner steel tube and actual failure occurs by failure of outer steel tube.

Table 2. Fatigue life cycles for IS, OS, and concrete.

Body	Fatigue life (Cycles)
Inner steel	8998
Concrete	634
Outer steel	1570
Fatigue life of CFDST	1570

2.3.1.2. Biaxiality indication

Biaxiality indication is defined as the principal stress smaller in magnitude divided by the larger principal stress with the principal stress nearest zero ignored. A biaxiality of zero corresponds to uniaxial stress, a value of −1 corresponds to pure shear, and a value of 1 corresponds to a pure biaxial state. The biaxiality indication of CFDST column is shown in Figure 3. It can be seen in the Biaxiality Figure 3, majority of this model has failed under a combination of axial stresses and shear stress.

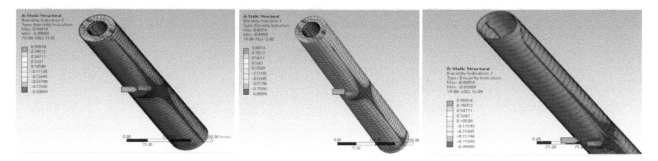

Figure 3. Biaxiality indication: (i) Outer steel (ii) Concrete (iii) Inner steel.

3. PARAMETRIC STUDY

The fatigue strength of materials is very sensitive to various external factors and internal factors. The influence of various factors on fatigue strength is an important aspect of fatigue research. Here, the main parameters studied are effect of axial force, drift, combined effect of force and drift, and loss of thickness of outer steel tube. The comparison between fatigue life of CFST and CFDST is also investigated. The parametric studies are conducted on CFDST column mentioned in Section 2.1. From Figure 4, it is visible that the inner steel tube is stressed below its capacity and it is surviving large cycles of fatigue, despite, it didn't improve the total fatigue life of the structure as the outer steel and concrete have already failed. The fraction of stress taken by the components in the ascending order is concrete, inner steel, and outer steel. That is, elements are stressed based on their strength and location from neutral axis.

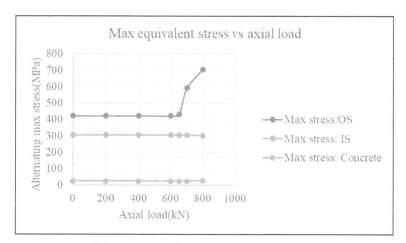

Figure 4. Equivalent stress on outer steel, inner steel, and concrete.

3.1 effect of drift on fatigue life

In order to study the effect of drift on fatigue strength of CFDST column, the CFDST column was checked for drift of 0.1%, 0.3%, 0.5%, 0.7% and 1% under constant axial force of 650kN ie., 77% of the ultimate capacity. From results as shown in Figure 6(a) and (b), it is clear that the stress on inner steel, outer steel, and concrete increases gradually up to percentage drift of 0.5%. Thereafter, the stresses showed a sudden increase and reached a constant value on reaching a drift of 0.8%. The number of cycles of life showed a gradual reduction from 15700 cycles to 1570 cycles on increasing drift from 0.1% to 0.5% as shown in Figure 7. Thereafter, it reduced suddenly to zero showing ultimate failure. The percentage reduction of life cycles can be seen in Figure 8. The CFDST column shows a percentage reduction of 93.4% on increasing drift from 0.1 to 0.5%, thereafter showing a sudden reduction to 100% on further increase of drift value to 0.7%. So, it can be concluded that considering the safety and serviceability of the structure, the maximum allowable drift under service load on a CFDST short column must be limited to 0.5%.

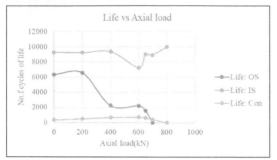

Figure 5. *(a) Fatigue life cycle of inner steel, concrete, and outer steel.*

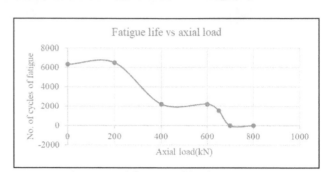

Figure 5. *(b) Fatigue life vs. axial load.*

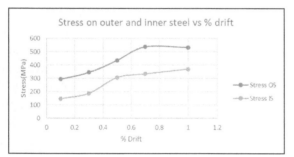

Figure 6. *(a) Stress on outer and inner steel vs. lateral displacement.*

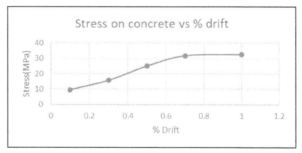

Figure 6. *(b) Stress on concrete vs. lateral displacement.*

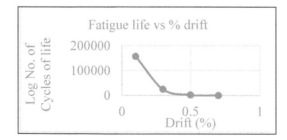

Figure 7. *Fatigue life vs drift*

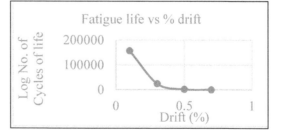

Figure 8. *Percentage change in life vs. drift.*

3.2 Effect of wall loss of outer steel on fatigue life

Wall loss is defined as the loss of material on the internal or external surface of steel tubes like pipelines, casing, or columns due to deteriorative effects of corrosion. Pitting, erosion, and uniform corrosion are the most common causes of wall loss. CFDST column was analyzed with different percentage loss of thickness of outer steel wall i.e., 1,10,20 and 33% reduction of outer steel thickness and the result of the analysis is described below. The percentage reduction in fatigue life is tabulated in Table 3. The stress on outer steel tube showed an increase due to reduction of outer steel tube thickness. Whereas the %increase in stress on concrete and inner steel is negligible. From Table 3, it is obvious that loss of thickness of outer steel tube causes failure of CFDST column when subjected to fatigue.

Table 3. Reduction in fatigue life.

Reduction in thickness of outer steel (%)	Reduction in fatigue life (%)
1	30.254
10	78.3653
20	81.469
33	100

3.4 Comparison between CFDST and CFST

In order to compare the fatigue strength of CFST and CFDST column, fatigue analysis was carried out under constant axial load of 650kN and drift of 0.5. The stresses on steel and concrete of CFST and CFDST are shown in Table 4. It is clear from Table 4, that, the stress on outer steel of CFST column is considerably high (18%) as compared to CFDST and the stress in concrete of CFST column is 28.4% higher than CFDST column. The fatigue life result shows that CFST didn't survive any fatigue cycles whereas CFDST survived 1570 cycles at 650kN axial load and 0.5% drift. Therefore, it can be concluded that the fatigue performance of CFST is weak compared to CFDST.

Table 4. Stress on outer steel, concrete, and inner steel.

Parameter	Body	CFST	CFDST	Variation (%)
Stress (MPa)	Outer steel	517.35	435.24	18
	Concrete	32.11	24.99	28.4
	Inner steel	–	305.86	–
Life (Cycles)	Outer steel	0	1570	100
	Concrete	0	634	100
	Inner steel		8998	-
Total Life (Cycles)		0	1570	100

4. CONCLUSIONS

1. Fatigue failure of CFDST column occurred at first on the concrete core due to combined bending and compressive force, nevertheless, the crushed concrete was contained in the double confinement of outer and inner steel tube and actual failure occurred by failure of outer steel tube.
2. Fatigue performance of CFDST column is maximum under an axial load of 70 to 77% ultimate capacity. The CFDST column cannot survive cycles of fatigue when loaded to its ultimate axial capacity.
3. The failure of CFDST under fatigue is abrupt on increasing the drift above the critical drift. So, considering the safety and serviceability of the structure, the maximum allowable drift under service load on a CFDST short column must be limited to 0.5%.

4. On comparing the fatigue performance of CFST and CFDST, it was found that the fatigue performance of CFST column is weak compared to the performance of CFDST column.

REFERENCES

[1] Uenaka, Hiroaki Kitoh, and Keiichiro Sonoda (2009). Concrete filled double skin circular stub columns under compression. Thin-Walled Structure, Elsevier, 48, February 2009.

[2] Pouria Ayough, N.H. Ramli Sulong, and Zainah Ibrahim (2019). Analysis and review of concrete-filled double skin steel tubes under compression. Thin-Walled Structures, Elsevier, November 2019.

[3] Ayoub Dehghani and Farhad Aslani (2019). A review on defects in steel offshore structures and developed strengthening techniques. Structures, Elsevier, 20, pp. 635-657, June 2019.

[4] Y. Essopjee and M. Dundu (2015). Performance of Concrete-Filled Double-Skin Circular Tubes in Compression. Composite Structures, Elsevier, 133, pp. 1276-1283, December 2015.

[5] R. Park and T. Paulay (1975). Reinforced Concrete Structures. Wiley Interscience Publication.

[6] M. Pagoulatou, T. Sheehan, X.H. Dai, and D. Lam (2014). Finite element analysis on the capacity of circular concrete-filled double-skin steel tubular (CFDST) stub columns. Engineering Structures, Elsevier, 72, pp. 102-112, April 2014.

[7] Tuan Trung Le, Vipulkumar Ishvarbhai Patel, Qing Quan Liang, and Phat Huynh (2021). Numerical modelling of rectangular concrete-filled double-skin steel tubular columns with outer stainless-steel skin. Journal of Constructional Steel Research, Elsevier, 179, January 2021.

32. Interfacial Bond Strength Between Normal and Geopolymer Concrete

Vishnu Sasidharan P,[1] Veena Nair A P,[2] and Job Thomas[3,*]

[1]Former M. Tech. Student, Department of Civil Engineering, School of Engineering, Cochin University of Science and Technology, Kochi, Kerala, India [2]M.Tech. Student, Department of Civil Engineering, School of Engineering, Cochin University of Science and Technology, Kochi, Kerala, India

[3]Professor, Department of Civil Engineering, School of Engineering, Cochin University of Science and Technology, Kochi, Kerala, India

*Corresponding author: job_thomas@cusat.ac.in,

ABSTRACT: The interfacial bond strength between two concrete surfaces was determined by conducting the slant shear test. The effects of the addition of steel fibers and recycled aggregate concrete on the compressive strength and slant shear strength of geopolymer concrete overlay were investigated. A total of 24 slant shear specimens with normal concrete as substrate and geopolymer concrete as an overlay and the companion cube specimens were cast and tested. The test results indicated that the addition of steel fibers increases the slant shear strength and the addition of recycled aggregates decreases the slant shear strength of geopolymer concrete. The prediction models for the compressive strength and slant shear strength are proposed.

KEYWORDS: Geopolymer concrete, recycled aggregate, slant shear strength, compressive strength, and steel fibers.

1. INTRODUCTION

With the tremendous growth of construction industries, the use of cement in concrete preparation is growing enormously. This will up a hazard to the environment due to excessive emission of greenhouse gases. Hence, it is important to find alternative binders to ordinary Portland cement (OPC). The term Geopolymer was coined by a French professor, Davidovitis in 1978 to represent a broad family of materials characterized by networks of inorganic molecules. A lot of research has been done on geopolymer concretes, which are made by combining industrial aluminosilicate waste materials with an alkaline solution, such as fly ash, GGBS, and metakaolin. Fly ash (FA) is a by-product of coal-fired power plants. Due to its widespread availability, useful silica ($SiO2$) and alumina-based composition, and lower water requirement, low-calcium fly ash is an ideal material for the production of geopolymers and can be used as a successful alternative to Portland cement. Whenever there is repair work in concrete, the shear transfer between the interface between the old substrate and the new layer of patching material is important. In this study, the shear strength of the joint between old and new layer of geopolymer concrete is studied.

Diab et al. (2017) studied the slant shear bond strength between self-compacting concrete and old concrete. The compressive strength of overlay, roughness of old concrete, addition of latex and use of polypropylene fiber were found to have a significant effect on the slant shear bond strength. Jafarinejad et al. (2019) observed that sand blasting method of surface preparation yields best interfacial bond shear strength fiber reinforced cement mortar and conventional concrete.

Liao et al. (2019) found that interfacial roughness treatment and increase in strength of mortar overlay improves the bond behaviour between the overlay and substrate. Feng et al. (2020) found that the carbon fibers can improve bond strength when strength of substrate and overlay are equal.

Peng et al. (2021) showed that the slant shear strength increases with surface roughness beyond 0.25 mm. At lower values of surface roughness, there is no trend could be found for slant shear strength. The earlier research indicates that the interfacial bond shear strength is an important property to be investigated. The investigations on the interfacial bond strength between geopolymer concrete overlay and conventional concrete substrate are limited. The present study attempts to fill this gap. In this study, the slant shear strength of geopolymer concrete is investigated and the influence of addition of steel fibers on bond strength is investigated.

2. EXPERIMENTAL STUDY

The interfacial bond shear strength was determined using slant shear test given by ASTM C882 (2020). The cylindrical specimens were cast with a shear angle of 30 degrees between the substrate and overlay and as shown in Fig. 1. The substrate was cast with conventional concrete with natural aggregate. Two strength grades of substrate concrete specimens, namely, 35 MPa and 55MPa, were prepared. The substrate specimens were cured for 28 days. Then the specimens were air dried for 2 months. Then the bonding surface was cleaned manually with a wire brush. The dust was removed by water washing and the surface is allowed to dry for 2 hours. The fresh geopolymer concrete was poured over the conventional concrete substrate. The variables considered in this study are replacement ratio of coarse aggregate, grade of geopolymer concrete, and steel fibers. Three specimens were cast in each set. The specimens were cured in ambient temperature for 7 days after casting the geopolymer concrete and tested. The average strength of three specimens was calculated and reported. The companion cubes were also cast and the strength was determined.

Figure 1. *Testing of Slant shear specimen*

Ordinary Portland cement of grade 53 was used for the preparation of substrate concrete. The specific gravity of the concrete was found to be 3.12. The crushed stone fine aggregate having a specific gravity of 2.73 was used. The natural coarse aggregates of nominal size 20 mm and 12.5 mm and having a specific gravity of 2.70 and 2.72 respectively were used. A superplasticizer containing sulfonated naphthalene formaldehyde was used. Micro silica having a specific gravity of 2.2 was used for preparing the concrete having a strength grade of 55MPa. The conventional concrete was designated by the strength grade. Conventional concrete designation M35 indicates the conventional concrete mix having a strength of 35 MPa on 28 days. The mix details of substrate concrete are given in Table 1.

Table 1. Mix details of substrate conventional concrete

Mix ID	Quantity of Constituent materials per cubic meter of concrete						
	Admixture	Water	Cement	Micro silica	Fine aggregate	Coarse aggregate (20 mm)	Coarse aggregate (12.5 mm)
M35	2.5	160	400	0	774	689	465
M55	2.9	162	450	45	733	652	438

The binder for geopolymer concrete was prepared by mixing class F fly ash and ground granulated blast furnace slag in a ratio 1:1. Recycled aggregates and hooked-end steel fibers were also used in the preparation of geopolymer concrete. The recycled aggregates prepared by crushing of concrete of a demolished bridge girder were used. The specific gravity of 20 mm and 12.5 mm nominal size recycled aggregate was 2.62 and 2.64 respectively. The steel fibers of length 60 mm length and 0.75 mm diameter glued in bundles were also used. The mix proportion of geopolymer concrete was designated by the strength grade, aggregate type, and fiber content. The geopolymer concrete designation 35R0 indicates that the strength grade is 35 MPa, aggregate type is recycled aggregate and the fiber content is 0 percent. The details of geopolymer concrete are given in Table 2.

Table 2. Details of geopolymer concrete mixes of overlay

Constituents	Quantity of constituent materials in kg per cubic meter of concrete for the mix designated by							
	35N0	35N1	35R0	35R1	55N0	55N1	55R0	55R1
Geopolymer binder	540	540	540	540	610	610	610	610
Fine aggregate	640	640	640	640	590	590	590	590
Natural coarse aggregate (20 mm)	560	560	0	0	520	520	0	0
Natural coarse aggregate (12.5 mm)	400	400	0	0	370	370	0	0
Recycled coarse aggregate (20 mm)	0	0	543	543	0	0	505	505
Recycled coarse aggregate (12.5 mm)	0	0	388	388	0	0	359	359
Alkali activator	360	360	360	360	410	360	410	410
Steel fibres	0	25	0	25	0	25	0	25

3. CALCULATIONS

The nominal strength of the specimen (σ_0) is calculated using Eq. (1)

$$\sigma_0 = \frac{P_u}{\left(\pi d^2 / 4\right)} \tag{1}$$

where P_u is the ultimate load and d is the diameter of the cylinder. The vertical stress (σ_0) can be resolved to shear (τ_α) and normal stress (σ_α) at the interface of the bonding surface inclined at an angle α and is given in Fig 2. The interfacial bond shear strength (τ_α) is calculated by

$$\tau_\alpha = \sigma_0 \cos\alpha \tag{2}$$

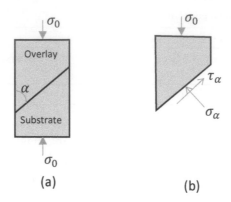

Figure 2. *Interfacial bond shear strength of slant shear specimen*

4. RESULTS AND DISCUSSIONS

A total of 24 specimens were prepared and tested. Three specimens corresponding to each specimen ID given in Table 3 were tested and average is reported. The companion cubes were also tested along with the slant shear specimen and the compressive strength of concrete were determined. The average compressive strength of three cubes was reported in Table 3.

Table 3. Results of compressive strength and slant shear strength of the specimens

Specimen ID	Overlay/ substrate	Mix designation	Compressive strength (MPa)	Slant shear strength, τ_α (MPa)
M35-35N0	Overlay	35N0	38.8	7.2
	Substrate	M35	36.1	
M35-35N1	Overlay	35N1	44.6	8.1
	Substrate	M35	36.1	
M35-35R0	Overlay	35R0	35.1	6.2
	Substrate	M35	36.1	
M35-35R1	Overlay	35R1	38.1	6.8
	Substrate	M35	36.1	
M55-55N0	Overlay	55N0	59.4	11.1
	Substrate	M55	56.6	
M55-55N1	Overlay	55N1	64.0	12.2
	Substrate	M55	56.6	
M55-55R0	Overlay	55R0	55.3	9.2
	Substrate	M55	56.6	
M55-55R1	Overlay	55R1	59.7	10.1
	Substrate	M55	56.6	

The compressive strength of geopolymer concrete is found to be increasing with the addition of steel fibers in the concrete. The increase in the compressive strength of 35N1 and 55N1 was found to be 14.9 % and 7.7 % respectively when compared to the corresponding control mixes 35N0 and 55N0. Similarly, the increase in the compressive strength was found to be 8% for both 35R1 and 55R1 mix when compared to the corresponding control mix 35R0 and 55R0. This may be due to the hoop confining stress developed due to the bridging of steel fibers, which prevents the lateral bulging of the specimen.

The compressive strength is found to decrease with the addition of recycled aggregate. The decrease in compressive strength for 35R0 and 55R0 was found to be 9.5 % and 6.9 % when compared to corresponding control mix 35N0 and 55N0 having natural aggregate. The presence of old mortar attached to the surface of the recycled aggregate will act as the weak spots in the interfacial zone between recycled aggregates and new mortar. This may be the reason for having lower strength in recycled aggregate geopolymer concrete when compared to natural aggregate geopolymer concrete.

The shear strength between the substrate conventional concrete and overlay geopolymer concrete is given in Table 3. The slant shear strength was found to increase with the addition of steel fibres and it was found to decrease with the addition of recycled aggregate to overlay. The increase in slant shear strength of 35N1 and 55N1 was found to be 12.5% and 9.9% when compared to the mixes of 35N0 and 55N0 having no fibers. Similarly, the increase in the slant shear strength was found to be 9.7 % for both of 35R1 and 55R1 when compared to the mixes 35R0 and 55R0. The presence of fibers at the interface mobilizes frictional force at the interface and this could be the reason for having higher interfacial shear strength for mixes having steel fibers.

The slant shear strength as found to decrease with the addition of recycled aggregate to concrete. The decrease in the slant shear strength of mixes 35R0 and 55R0 was found to be 13.8% and 17.1% when compared to 35N0 and 55N0. The weak zones due to the presence of old mortar at the surface of recycled aggregate might reduce the efficiency of load redistribution within the composite matrix. This is the reason for having lower slant shear strength in recycled aggregate geopolymer concrete when compared to the natural aggregate geopolymer concrete overlay.

5. PREDICTION MODELS

The prediction models for the compressive strength and the slant shear strength of geopolymer concrete were developed by regression analysis and are given in Table 4. The dependent variables of the prediction model are the compressive strength of natural aggregate geopolymer concrete (f_{cu}), replacement ratio of recycled coarse aggregate in percentage (R_{CA}) and fiber percentage by weight (W_f).

Table 4. Prediction model for strength parameters

Strength property	Prediction model	R- square value
Compressive strength (MPa)	$1.006 f_{cu} - 0.046 R_{CA} + 4.507 W_f$	0.99
Slant shear strength (MPa)	$0.186 f_{cu} - 0.015 R_{CA} + 0.967 W_f$	0.99

The regression coefficient of R_{CA} is found to be negative, which indicates that the strength properties reduce with an increase in R_{CA}. Similarly, the regression coefficient of W_f is found to be positive, which indicates the strength properties increase with the increase in W_f. The variation of R_{CA} and W_f in the prediction model corroborates with the experimental results. The R-square value of the prediction models is found to be 0.99, which indicates that the models are capable of predicting 99 percent variation of the experimental results.

6. CONCLUSIONS

Based on the experimental study carried out, the following conclusions are arrived at

- The compressive strength and slant shear strength of geopolymer concrete increase with the addition of steel fibers
- The addition of recycled aggregate concrete reduces the compressive and slant shear strength of the geopolymer concrete

REFERENCES

[1] Diab, A. M., Abd-Elmoaty, A. E. M., and Eldin, M. R. T. (2017). Slant shear bond strength between self-compacting concrete and old concrete. Construction and Building Materials, 130, 73–82.

[2] Jafarinejad, S., Rabiee, A., and Shekarchi, M. (2019). Experimental investigation on the bond strength between Ultra high strength Fiber Reinforced Cementitious Mortar and conventional concrete. Construction and Building Materials, 229, https://doi.org/10.1016/j.conbuildmat.2019.116814.

[3] Liao, W., Wang, H., Li, M., Ma, C., and Wang, B. (2019). Large scale experimental study on bond behavior between polymer modified cement mortar layer and concrete. Construction and Building Materials, 228, https://doi.org/10.1016/j.conbuildmat.2019.116751.

[4] Feng, S., Xiao, H., Zhang, R., and Yang, C. (2020). Bond performance between substrate concrete and repair mortar: Effect of carbon fibre and expansive agent. Construction and Building Materials, 250, https://doi.org/10.1016/j.conbuildmat.2020.118830.

[5] Peng, G., Niu, D., Hu, X., Pan, B., and Zhong, S. (2021). Experimental study of the interfacial bond strength between cementitious grout and normal concrete substrate. Construction and Building Materials, 273, https://doi.org/10.1016/j.conbuildmat.2020.122057.

[6] ASTM C882 (2020). Standard test method for bond strength of epoxy-resin systems used with concrete by slant shear. ASTM International, USA.

33. Properties of Geopolymer Concrete-Containing Hybrid Length Steel Fibers

Mathukutty Sebastian, Geever Ambadan, and Job Thomas*
Department of Civil Engineering, School of Engineering, Cochin University of Science and Technology, Kochi, Kerala 682022, India
*Corresponding author: job_thomas@cusat.ac.in

ABSTRACT: Geopolymer concrete is a quasi-brittle material. The steel fibers are used to mitigate the brittle behavior of conventional concrete. In this study, the effect of the addition of steel fibers on the strength properties of geopolymer concrete is investigated. Hooked-end macro and straight micro steel fibers are added to the geopolymer concrete. The compressive strength, modulus of elasticity, splitting tensile strength, modulus of rupture, pull-out strength, and pull-off strength of geopolymer concrete were determined. All the strength properties were found to increase with the addition of steel fibers. The strength prediction models based on regression analysis are proposed. The proposed prediction models can be used for the designer to finalize the steel fiber dosage for getting the desired magnitude of the strength property.

KEYWORDS: Geopolymer Concrete, Steel Fiber, Compressive Strength, Tensile Strength, Pull-Out Strength, Pull-Off Strength

1. INTRODUCTION

Environmental protection has become an important criterion for public policy, with the primary objective being to reduce emissions of greenhouse gases. The cement industry contributes 5% of global CO_2 emissions, and the use of materials with lower environmental impact has gained traction. However, using pozzolanic materials such as fly ash and ground granulated blast furnace slag (GGBS) can be used in the construction industry to reduce carbon emissions. Fiber reinforcement is yet another additive that can reduce brittleness and crack propagation in concrete. In this study, geopolymer concrete containing fly ash, GGBS, and fibers was prepared and the mechanical properties were determined.

Numerous studies on FRC describe that the efficacy of concrete composites will improve when two or more fiber types are combined. Ganesh and Muthukannan (2021) found that steel fibers can improve tensile parameters, while glass fibers decreased workability. The compressive, split tensile, and flexural strength of geopolymer concrete is found to increase with the addition of steel fibers. Rahman and Al-Ameri (2021) developed a new proportion of self-compacting geopolymer concrete (SCGC) with reduced alkali activators and no super plasticizers. It is found that the fly ash/slag ratio affects setting time when water/binder ratio is maintained at 0.45. The average compressive strength of SCGC is found to increase with age.

Al-Majidi et al. (2017) studied the strength of plain and steel fiber reinforced geopolymer concrete (SFRGC) containing silica fume (SF) and GGBS. Steel fiber addition was found to reduce compressive strength but found to improve post-crack load-carrying capacity. Wang et al. (2021) investigated the influence of steel fibers on the flowing ability, compressive strength, and flexural behavior of ultra-high performance geopolymer concrete (UHPGC). The incorporation of 1% steel fibers increases compressive and flexural strengths, while the increase of fiber diameter reduces compressive and flexural strengths. The reinforcing efficiency of fibers is found to influence by the aspect ratio and size of fiber. Zhang et al.

Chapter 33 DOI- 10.1201/9781032657271-33

(2021) proposed that bond performance between steel rebars and concrete can be improved by aligning limited steel fibers around the magnetized rebar.

Geopolymer concrete reduces CO_2 emissions by up to 45% while staying economically competitive, making it a sustainable, ecologically friendly alternative to Ordinary Portland Cement (OPC)-based concrete. In comparison to conventional cement concrete, steel fiber-reinforced geopolymer concrete has higher early strength and durability, making it a suitable replacement in the precast industry. Future shortages of cement for the manufacturing of concrete can be effectively tackled.

2. EXPERIMENTAL STUDY

The materials such as geopolymer binder, geo-activator, fine aggregate, coarse aggregate, and macro and micro steel fiber were used in this study. The geopolymer binder is prepared by adding fly ash and GGBS in a proportion of 50:50 low calcium fly ash of class F was used. The alkaline activator concentration is the most critical factor for successful geopolymer formation and the evolution of high compressive strength. Geopolymer cement is normally used in the preparation of geopolymer concrete. Crushed stones of 12.5 and 20 mm in size and having a specific gravity of 2.69 and 2.65 were used in this study. Crushed stone sand having a specific gravity of 2.69 was used as fine aggregate in this study. Hooked-end steel fibers of 30 mm long and 0.38 mm diameter were used as macro fibers and straight steel fibers of 20 mm long and 0.22 mm diameter were used as micro fibers. Steel fibers were randomly dispersed in concrete. The compressive strength, modulus of elasticity, and bond strength were determined in the hardened stage. A total of six concrete mixes were prepared. The mix was designated to match this grade of concrete and the percentage of macro and micro steel fiber. M60/1/0 indicates M60 grade concrete having 1% macro steel fiber and 0% micro steel fiber by weight. The details of weight of the constituent materials used for the preparation of concrete mixes are given in Table 1.

Table 1. Proportion of constituent materials.

Constituent materials	Weight of constituent materials in kg/m³ of concrete for the mix designations					
	M35/0/0	M35/1/0	M35/1/2	M60/0/0	M60/1/0	M60/1/2
GGBS	260	260	260	325	325	325
Fly ash	260	260	260	325	325	325
Activator	363	363	363	412	412	412
Fine aggregate	660	660	660	578	578	578
Coarse aggregate (20 mm)	602	602	602	542	542	542
Coarse aggregate (12 mm)	401	401	401	361	361	361
Macro fiber (30 mm)	-	25.50	25.50	-	25.50	25.50
Micro fiber (12 mm)	-	-	51	-	-	51

The compressive strength of concrete was determined using standard cube specimens as given by the methods specified in IS: 516 (1959). Modulus of elasticity of concrete is also determined using cylinder specimen and extensometer (IS: 516, 1959). The splitting tensile strength of concrete is determined using the standard cylinder specimen as given in IS:5816 (1999). The modulus of rupture of concrete is determined by flexure test as given by IS:516 (1959). The pull-out test is conducted as per BS EN 12504-3 (2005) to determine the strength of the concrete when subjected to pull-out force. A hole was drilled into hardened concrete and an oversized screw was inserted into the hole. The screw was subsequently pulled out from the concrete in pull out test. The pull-off test was conducted as given by BS EN 1542-1 (1999). A metallic disc is bonded to the surface of this concrete. The disc is then pulled to detach from this specimen, which is taken as a measure of strength of concrete in pull-off test. Three specimens were tested and average of the three test results was determined.

3. RESULTS AND DISCUSSIONS

A total of 36 specimens were cast and tested to determine the compressive strength, modulus of elasticity, splitting tensile strength, modulus of rupture, pull-out strength, and pull-off strength. Average of three test data is reported in Table 2.

Table 2. Mechanical strength properties of geopolymer concrete.

Mix designation	Compressive strength (MPa)	Modulus of elasticity (GPa)	Splitting tensile strength (MPa)	Modulus of rupture (MPa)	Pull-out strength (MPa)	Pull-off strength (MPa)
M35/0/0	38.58	22.0	2.47	3.07	0.70	2.45
M35/1/0	44.58	23.1	2.81	3.57	1.09	3.20
M35/1/2	47.36	27.0	2.98	3.81	1.14	3.68
M60/0/0	63.50	30.2	3.40	4.26	1.35	2.75
M60/1/0	68.04	31.0	3.92	4.90	1.92	3.45
M60/1/2	71.03	35.1	4.16	5.20	2.29	3.80

The compressive strength was found to increase with an increase in fiber content in M35 and M60 grade concrete. The increase in compressive strength for M35/1/2 was found to be 22% when compared to this corresponding control mix M35/0/0. The modulus of elasticity under compression of cement was also found to increase fiber content. The increase in modulus of elasticity of mix M60//2 is found to be 16% when compressed to that of corresponding control mix M60/0/0. This may be due to the fact that the fibers bridging across the cracks induces hydrostatic pressure which eventually leads to confinement in concrete.

The splitting tensile strength and modulus of rupture were also found to increase with the increase in fiber content. The increase in splitting tensile strength of M60/1/2 is found to be 22% when compared to M60/0/0. Similarly, the increase in modulus of rupture of mix M35/1/2 is found to be 24% when compared to M35/0/0. This indicates that fibers bridging across the crack contribute significantly to the tensile strength of this concrete.

The pull-out strength and pull-off strength of concrete significantly increase with the addition of macro and microfibers. The increase in the pull-out strength for mix M35/1/2 is found to be 62% when compared to that of mix M35/0/0. Similarly, the increase in the pull-off strength of mix M60/1/2 is found to be 38% when compared to that of mix M60/0/0. This indicates that fibers bridging across the crack formed during pull-out and pull-off test mobilize the resistance against further cracking.

4. PREDICTION OF STRENGTH PROPERTIES

The regression perdition model for the strength properties of geopolymer hybrid length steel fiber-reinforced concrete was proposed. Table 3. The dependent variables are the strength grade of concrete (f_{ck}), percentage weight fraction of macro fibers (W_{fM}), and percentage weight fraction of micro fibers (W_{fm}). In all models, the regression coefficient for weight fraction of micro and macro fibers is found to be positive, which indicates that the addition of steel fibers is beneficial to improve the strength properties of the geopolymer concrete. The R-square value of the multiregression analysis model is found to be greater than 96%. Hence, it can be concluded that the variation in the data is represented by the proposed model quite accurately.

Table 3. Prediction model for strength properties.

Strength property	Prediction model	R²
Compressive strength (MPa)	$f_{cu} = 0.960 f_{ck} + 5.27 W_{fM} + 1.44 W_{fm} + 5.40$	0.99
Modulus of elasticity (GPa)	$E_{cf} = 4.408\sqrt{f_{ck}} + 0.95 W_{fM} + 2.00 W_{fm} - 4.01$	0.99
Splitting tensile strength (MPa)	$f_{sp} = 0.586\sqrt{f_{ck}} + 0.43 W_{fM} + 0.10 W_{fm} - 1.07$	0.99
Modulus of rupture (MPa)	$f_{r} = 0.712\sqrt{f_{ck}} + 0.57 W_{fM} + 0.13 W_{fm} - 1.20$	0.99
Pull-out strength (MPa)	$f_{pout} = 0.479\sqrt{f_{ck}} + 0.48 W_{fM} + 0.10 W_{fm} - 2.24$	0.96
Pull-off strength (MPa)	$f_{poff} = 0.122\sqrt{f_{ck}} + 0.72 W_{fM} + 0.20 W_{fm} + 1.76$	0.99

5. CONCLUSIONS

Based on the experimental study carried out on steel fiber reinforced geopolymer concrete, the following conclusions are arrived at:

- The addition of steel fibers increases the compressive, tensile, and pull strength of geopolymer concrete.
- The addition of steel fibers is beneficial both in normal and moderately high-strength grades of geopolymer concrete.
- The pull-out and pull-off strengths of geopolymer concrete also increase with the addition of steel fibers.
- The variation in the experimental data is well represented in the prediction model proposed in this study.

The models proposed will be useful for the designers to compute the required fiber volumes to be incorporated corresponding to a target strength value.

REFERENCES

[1] Al-Majidi, M. H., Lampropoulos, A., and Cundy, A. B., "Tensile properties of a novel fibre reinforced geopolymer composite with enhanced strain hardening characteristics," *Comp Stru*, vol. 168, pp. 402–427, 2017.

[2] BS EN 12504-3, *Testing Concrete in Structures*, British Standards Institution, 2005.

[3] BS EN 1542-1, *Products and Systems for the Protection and Repair of Concrete Structures – Test Methods – Measurement of Bond Strength by Pull-Off*. British Standards Institution, 1999.

[4] Ganesh, A. C., and Muthukannan, M., "Development of high performance sustainable optimized fibre reinforced geopolymer concrete and prediction of compressive strength," *J Clean Prod*, vol. 282, 2021. doi: 10.1016/j.jclepro.2020.124543.

[5] IS 5816, *Splitting Tensile Strength of Concrete – Method of Test*, New Delhi: Bureau of Indian Standards, 1999.

[6] IS: 516, *Methods of Tests for Strength of Concrete*, New Delhi: Bureau of Indian Standards, 1959.

[7] Rahman, S. K., and Al-Ameri, R., "A newly developed self-compacting geopolymer concrete under ambient condition," *Const Build Mat*, vol. 267, 2021. doi: 10.1016/j.conbuildmat.2020.121822.

[8] Wang, Y. S., Peng, K. D., Alrefaei, Y., and Dai, J. G., "The bond between geopolymer repair mortars and OPC concrete substrate: Strength and microscopic interactions," *Cem Conc Comp*, vol. 119, 2021. doi: 10.1016/j.cemconcomp.2021.103991.

[9] Zhang, W., Lee, D., Lee, C., Zhang, X., and Ikechukwu, O. "Bond performance of SFRC considering random distributions of aggregates and steel fibers," *Const Build Mat*, vol. 291, 2021. doi: 10.1016/j.conbuildmat.2021.123304

34. Flexural Strength of Geopolymer Concrete-Reinforced Beams Containing Steel Fibers

Rajesh G Nair, Nikita Rechal, Fabitha K J and Job Thomas*
Department of Civil Engineering, School of Engineering, Cochin University of Science and Technology, Kochi, Kerala 682022, India
*Corresponding author: job_thomas@cusat.ac.in

ABSTRACT: The effect of the addition of steel fibers on reinforced concrete beam made up of geopolymer concrete was studied. The variables of the study are the grade of concrete, reinforcement ratio, and steel fiber content. The geopolymer concretes of strength grade 25 and 35 MPa were used. Recycled coarse aggregate was used for the preparation of the geopolymer concrete. Two types of steel fibers, namely, micro straight filament type and macro hooked end type were used. The test results indicated that the flexural strength of geopolymer concrete reinforced beams increases with the concrete strength, longitudinal reinforcement ratio, and steel fiber content. The prediction models suggested in this study corroborate with the experimental data.

KEYWORDS: Geopolymer-Reinforced Concrete, Steel Fibers, Compressive Strength, Tensile Strength

1. INTRODUCTION

The sustainable development in the construction industry includes the use of non-conventional and innovative materials, and the recycling of waste materials. This will help to minimize the depletion of natural materials and conserve the environment. The recycling of demolition wastes to produce aggregates is a greener approach. The use of these materials in structural systems will have to be explored. The behavior of recycled aggregate in geopolymer concrete in beams needs to be studied. In this study, the flexural strength of reinforced geopolymer concrete beams containing recycled aggregate and steel fibers is explored.

Bhuttaet al. (2016) found that the flexural strength of geopolymer matrix increases with the addition of steel fibers. Kathirvel and Kaliyaperumal (2016) concluded that the deformation of recycled aggregate concrete beams is more than that of the corresponding normal aggregate concrete beams. Gao and Zhang (2016) observed that the addition of steel fibers up to a volume fraction of 2% is beneficial to increase in flexural strength of reinforced concrete beams containing recycled aggregates. Tran et al. (2019) reported that there is a reduction in the flexural strength of the geopolymer concrete beam when 1.5% by volume of 35 mm long steel fibers are added. It was explained that this reduction may be due to the poor distribution of steel fibers in the mix. El-Sayedand Shaheen (2020) showed that the flexural strength of geopolymer concrete beam containing lathe waste increases with the addition of steel fibers. Based on the review of the literature, it is understood that the detailed experimental study of flexural performance of recycled aggregate geopolymer concrete beams is not reported. In this study, the effect of the addition of steel fibers on the flexural strength of recycled aggregate concrete beam is investigated.

2. MATERIALS AND METHODS

Binders, alkali activators, fine aggregate, and coarse aggregate were used for the preparation of geopolymer concrete. The binder consisting of fly ash and ground granulated blast furnace slag in the ratio 1:1 by weight was used. The mixture of sodium hydroxide (NaOH) and sodium silicate (Na2SiO3) in the ratio of 1:2.5 by weight was used as alkali activator. The crushed granite fines conforming to Zone II of IS:383 (2016) were used as fine aggregate. The recycled aggregate of nominal size 20 and 12.5 mm obtained by crushing the concrete of abandoned bridge project was used as coarse aggregate. Micro and macro steel fibers were used. Straight filament-type brass coated steel fibers of 6 mm long and 0.22 mm diameter were used as micro steel fibers. Macro steel fibers of hooked end type 30 mm long and 0.38 mm diameter bundled together with water-soluble glue were used as macro steel fiber. The mixes were designated to indicate the strength grade and the percentage by weight of micro and macrofibers. For example, 35-1-2 indicates the mix of strength grade 35 MPa containing 1% of microfiber and 2% of macrofiber. The mix proportion of various mixes is given in Table 1.

Table 1. Details of geopolymer concrete.

Constituents	Quantity of constituent materials in kg/m³ of concrete for the mix designated by					
	35-0-0	35-1-0	35-1-2	55-0-0	55-1-0	55-1-2
Geopolymer binder	540	540	540	610	610	610
Alkali activator	360	360	360	410	410	410
Fine aggregate	640	640	640	590	590	590
Recycled coarse aggregate (20 mm)	543	543	543	505	505	505
Recycled coarse aggregate (12.5 mm)	388	388	388	359	359	359
Micro steel fiber (6 mm)	0	25	25	0	25	25
Macro steel fiber (30 mm)	0		50	0	0	50

Steel reinforcement of grade 415 MPa was used as longitudinal and transverse steel in the beam. Beams having different reinforcement ratios were cast and details are given in Figure 1(a)–(d). In all beams, two-legged stirrups of 8 mm diameter were provided at a spacing of 100 mm center to center. A total of 12 beams having an overall dimension 200 mm × 300 mm × 2200 mm were cast. The effective depth of the longitudinal reinforcement was 250 mm in all beams. These beams were tested over a span of 2000 mm. The beams were designated to indicate the mix details and the reinforcement ratio. For example, 35-0-0/0.804 indicates the beam with mix 35-0-0 and having a longitudinal reinforcement ratio of 0.804%. The beams were placed in ambient conditions for seven days and then tested. Three cubes of size 150 mm × 150 mm × 150 mm were along with the beams to determine the compressive strength of the concrete. The companion cubes cast along with the beams were tested on the same day as that of the beam.

3. CALCULATION

The ultimate moment (M_u) of beam is calculated by

$$M_u = P_u a \tag{1}$$

where P_u is the ultimate load of the beam obtained in the experiment and a is the shear span of the beam. In this study, is equal to 0.65 m Similarly, moment of resistance of the beam is calculated by Eq. (2), which is the method suggested by IS:456 (2000) for normal concrete beam.

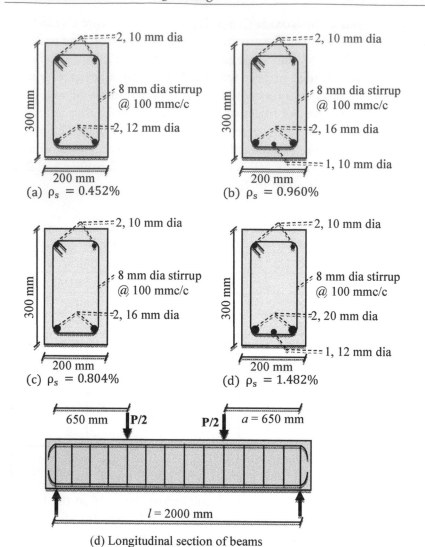

(a) $\rho_s = 0.452\%$

(b) $\rho_s = 0.960\%$

(c) $\rho_s = 0.804\%$

(d) $\rho_s = 1.482\%$

(d) Longitudinal section of beams

Sl. No.	Beam designation
1	25-0-0/0.452
2	25-1-0/0.452
3	25-1-2/0.452
4	25-0-0/0.960
5	25-1-0/0.960
6	25-1-2/0.960
7	35-0-0/0.804
8	35-1-0/0.804
9	35-1-2/0.804
10	35-0-0/1.482
11	35-1-0/1.482
12	35-1-2/1.482

Figure 1. *Details of the beam.*

$$MR_l = 0.138 f_{ck} b d^2 \tag{2}$$

where f_{ck} is the grade of concrete, b is the width of the beam, and d is the depth of the beam. The limiting value of the moment of resistance (MR_l) given by Eq. (2) is valid for Fe 415 grade steel. MR_l corresponds to a balanced section, in which the quantity of steel in singly reinforced section is sufficient to develop yielding strain in steel (0.0038 in Fe415 grade steel) and maximum strain (0.0035) in concrete simultaneously. The moment of resistance (MR) of the section calculated by

$$MR = 0.87 f_y \left(\frac{p_t}{100} \right) \left(1 - 1.005 \left(\frac{f_y}{f_{ck}} \right) \left(\frac{p_t}{100} \right) \right) b d^2 \tag{3}$$

where tensile reinforcement percentage (p_t) is given by

$$p_t = \left(\frac{A_{st}}{bd} \right) 100 \qquad \sim \tag{4}$$

where A_{st} is the area of tensile steel in the beam. The permissible value of moment of resistance (MR_p) is calculated by minimum of MR_l and MR.

$$MR_p = MINIMUM\left(MR, MR_l\right) \tag{5}$$

MR_l and MR are originally developed for normal concrete containing natural aggregate and no steel fibers.

4. RESULTS AND DISCUSSIONS

The flexural strength of the test beam is reported in Table 2. The companion cubes were tested and the average strength of geopolymer concrete is also given in Table 2.

Table 2. Strength of concrete and test beams.

Sl. no.	Beam designation	Compressive strength of concrete, f_{cu} (MPa)	Ultimate load, P_u (kN)	Ultimate moment, M_u (kNm)	Moment of resistance, MR_p (kNm)
1	25-0-0/0.452	25.1	39.6	25.7	18.9
2	25-1-0/0.452	27.2	43.8	28.5	19.0
3	25-1-2/0.452	29.2	48.7	31.7	19.1
4	25-0-0/0.960	25.3	60.2	39.1	36.5
5	25-1-0/0.960	26.9	67.1	43.6	36.9
6	25-1-2/0.960	28.9	73.6	47.8	37.3
7	35-0-0/0.804	35.6	56.4	36.7	32.9
8	35-1-0/0.804	37.4	62.3	40.5	33.0
9	35-1-2/0.804	40.1	69.8	45.4	33.3
10	35-0-0/1.482	36.1	96.1	62.5	55.4
11	35-1-0/1.482	38.1	108.2	70.3	56.0
12	35-1-2/1.482	40.8	118.1	76.8	56.8

The compressive strength of geopolymer concrete was found to increase with the increase in steel fibers. The average increase in the compressive strength of concrete mix 25-1-2 was found to be 15.2% when compared to 25-0-0. Similarly, the increase in the compressive strength of mix 35-1-2 was found to be 12.8% when compared to the mix 35-0-0. The increase in compressive strength by adding steel fibers can be attributed to the confining stress offered by steel fibers bridging across the vertical peripheral cracks in the test specimen.

The ultimate flexural load of the beam was found to increase with the increase in steel fibers. The increase in flexural load in beams 25-1-2/0.452 and 25-1-2/0.960 was found to be 23.0% and 22.3% when compared to the corresponding control beams 25-0-0/0.452 and 25-0-0/0.960. Similarly, the increase in the flexural load in beams 35-1-2/0.804 and 35-1-2/1.482 was found to be 23.8% and 22.9%, respectively, when compared to the corresponding control beams 35-0-0/0.804 and 35-0-0/1.482. Similar to longitudinal reinforcement, steel fibers bridge across flexural crack in the beam and will act as supplementary reinforcement preventing the opening of crack. This enhances the resistance against flexural load in the beams containing steel fibers. This is the reason for having higher strength in beams containing steel fibers when compared to the corresponding control beams having no fibers.

The test results indicated that the flexural strength increases with increase in concrete strength and longitudinal steel in the beams. This is because the moment of resistance in a beam is mobilized by the compressive force offered by the concrete above the neutral axis and tensile force below the neutral axis. The ultimate moment of the cross-section was also calculated using Eq. (1). The ultimate moment is directly proportional to the applied load and hence the variation in the ultimate moment was found to be the same as that of the ultimate load of the respective beam. The moment of resistance (MR_p) of the beam is calculated based on the equation given by IS:456 (2000). The effect of fibers bridging across the crack below the neutral axis was not accounted for in the model given by IS:456 (2000). Hence, the benefit of adding steel fibers in beams was not reflection in the moment of resistance value predicted by IS:456 (2000). The difference between the magnitude of moment of resistance predicted for the control beam and the beams containing steel fibers was found to be small. This indicates that the effect of the presence of fibers required to be incorporated separately.

5. PREDICTION MODEL

The prediction model for the compressive strength of geopolymer concrete is proposed based on regression analysis. Similarly, the prediction model for the ultimate moment of geopolymer concrete beams is also proposed and is given in Table 3.

Table 3. Proposed prediction models for fiber-reinforced geopolymer concrete.

Property	Prediction model	R^2
Compressive strength (MPa)	$f_{cuc} + 1.811W_{fm} + 1.175W_{fM}$	0.99
Ultimate moment (kNm)	$1.15MR_{pc} + \left(94.40W_{fm} + 47.98W_{fM}\right)\left(D - x_u\right)b$	0.99

where f_{cuc} is the compressive strength of control concrete having no fibers and MR_{pc} is the permissible moment of resistance of control beam having no fibers. W_{fm} and W_{fM} are the percentage weight of micro and macro steel fibers in the concrete. x_u is the depth of the neutral axis and is calculated by Eq. (6). The multiplication term $\left(D - x_u\right)b$ represents the cross-sectional area of the beam below the neutral axis.

$$x_u = \left(\frac{0.87 f_y A_{st}}{0.36 f_{ck} b}\right) \tag{6}$$

The regression coefficient of the terms W_{fm} and W_{fM} are found to be positive. This means that there will be an increase in the compressive strength and ultimate moment with the addition of steel fibers. This corroborates with the experimental data. The R^2 value of the prediction model is 0.99, which indicates that the variation in the experimental data is well predicted by the model.

6. CONCLUSIONS

Based on the experimental study on the flexural strength of reinforced geopolymer concrete beams containing steel fibers, the following conclusions were arrived at:
- The compressive strength of geopolymer concrete increases with the addition of steel fibers.
- The increase in the flexural strength of geopolymer concrete beam is significant when steel fibers are added.
- The flexural strength of geopolymer concrete increases with the increase in longitudinal reinforcement and grade of concrete.
- The prediction based on the proposed models accounts for the variation in the experimental data.

REFERENCES

[1] Bhutta, A., Ribeiro Borges, P. H., Zanotti, C., Farooq, M., and Banthia, N., "Flexural behaviour of geopolymer composites reinforced with steel and polypropylene macro fibers," *Cem Conc Comp*, vol. 80, pp. 31–40, 2016.

[2] El-Sayed, T. A., and Shaheen, Y. B. I., "Flexural performance of recycled wheat straw ash-based geopolymer RC beams and containing recycled steel fibre," *Structures*, vol. 28, pp. 1713–1728, 2020.

[3] Gao, D., and Zhang, L., "Flexural performance and evaluation method of steel fibre reinforced recycled coarse aggregate concrete," *Constr Build Mat*, vol. 159, pp. 126–136, 2018.

[4] IS:383, *Coarse and Fine Aggregate for Concrete – Specification*, New Delhi: Bureau of Indian Standards, 2016.

[5] IS:456, *Plain and Reinforced Concrete – Code of Practice*, New Delhi: Bureau of Indian Standards, 2000.

[6] Kathirvel, P., and Kaliyaperumal, S. R. M., "Influence of recycled concrete aggregates on the flexural properties of reinforced alkali activated slag concrete," *Const Build Mat*, vol. 102, pp. 51–58, 2016.

[7] Tran, T. T., Pham, T. M., and Hao, H., "Experimental and analytical investigation on flexural behaviour of ambient cured geopolymer concrete beams reinforced with steel fibres," vol. 200, 2019. doi: 109707.

35. Properties of Recycled Aggregate Concrete-Containing Steel Fibers

Joby C M, Soniya Thomas, and Job Thomas*
Department of Civil Engineering, School of Engineering, Cochin University of Science and Technology, Kochi, Kerala 682022, India
*Corresponding author: job_thomas@cusat.ac.in

ABSTRACT: Use of recycled aggregate in the concrete is important to reduce the excessive mining in the natural granite quarries. In this study, the strength properties such as compressive strength, split tensile strength, modulus of rupture, modulus of elasticity, and rebar pull-out bond shear strength of recycled aggregate concrete are determined experimentally. The effect of the addition of steel fibers in recycled aggregate concrete is also studied. The variables of the study are the grade of concrete and the percentage weight fraction of steel fibers. The concrete having a compressive strength of 30 and 50 MPa is prepared. Two types of steel fibers, namely, hooked-end macro steel fiber and brass-coated straight filament type micro steel fiber are used. The test results indicate that the addition of steel fibers is beneficial in improving the properties of both grades of concrete. The prediction models for the various strength properties are proposed in this study.

KEYWORDS: Recycled Aggregate Concrete, Compressive Strength, Split Tensile Strength, Modulus of Rupture, Modulus of Elasticity, Bond Strength, Prediction Model, Steel Fibers

1. INTRODUCTION

The construction industry in India is experiencing rapid urbanization, which comes with a moral responsibility of protecting the environment. To achieve this, major emphasis is to be laid on the use of wastes and by-products in new constructions. This includes slag, power plant wastes, recycled wastes, mining and quarrying wastes, waste glass, incinerator residue, red mud, burnt clay, sawdust, combustor ash, and foundry sand. The crushed pieces of concrete from old buildings, concrete pavements, bridges, and other structures exceeded their age can be used as aggregates in concrete. It will help in reducing the excessive mining of natural rock and save the environment.Mi et al. (2020) found that the compressive strength of concrete containing recycled aggregate is 13% more than the natural aggregate concrete. It was suggested that the presence of dense mortar of old concrete on the surface of recycled aggregate might have helped to improve the strength characteristics of concrete containing recycled aggregate. Thomas et al. (2018) observed that the reduction in compressive strength of concrete containing recycled aggregate is 34% when compared to the concrete containing natural aggregate. Dimitriou et al. (2018) reported that the reduction in compressive strength is 25% and 34%, respectively, when 50 and 100 percentage of natural aggregates in concrete is replaced by recycled aggregate. Anike et al. (2020) observed that the increase in the split tensile strength is 4% when 1.5% (by volume) of steel fibers are added to the recycled aggregate concrete. Gao et al. (2020) showed that the durability properties of recycled aggregate concrete can be improved by adding steel fibers to it.The effect of the addition of micro and macro fibers in recycled aggregate concrete has not been studied earlier. This study addresses this lacuna. In this study, the compressive strength, tensile strength, modulus of elasticity, and rebar pull-out strength of recycled aggregate concrete containing micro and macro steel fibers were determined experimentally.

Chapter 35 DOI- 10.1201/9781032657271-35

2. MATERIALS AND METHODS

Ordinary Portland cement of 53 grade confirming to IS:12262 (1987) was used in the preparation of concrete. The crushed stone sand of specific gravity 2.72 and confirming to zone II of IS:383 (2016) was used as fine aggregate. Natural crushed stones and recycled aggregate from concrete were used as coarse aggregate. The specific gravity of 20 mm and 12.5 mm nominal size recycled aggregate was found to be 2.60 and 2.63, respectively. Silica fumes of specific gravity 2.20 was used. Two types of fibers steel were used, namely, micro fibers and macro fibers. Straight brass coated wire fibers having a length of 6 mm and a diameter of 0.22 mm were used as microfibers and hooked-end fibers having a length of 30 mm and a diameter of 0.38 mm were used as macro fiber.

Recycled aggregates were used for the making of concrete. The mix proportion for the concrete was designed based on the absolute volume method as given in IS:10262 (2009). Two strength grades were prepared, namely 30 and 50 MPa. The steel fibers were added by the weight of the concrete mix. The mixes were designated to indicate grade and weight fraction of micro and macro fiber. For example, 30-1-2 indicates the concrete mix of 30 MPa strength containing 1 percentage micro fiber and 2 percentage macro fiber. The details of the mix proportion are given in Table 1.

Table 1. Concrete mix for recycled aggregate concrete.

Constituent materials	Quantity in kg/m³ of concrete for mix designation					
	30-0-0	30-1-0	30-1-2	50-0-0	50-1-0	50-1-2
Cement	400	400	400	450	450	450
silica fume	-	-	-	45	45	45
Fine aggregate	770	770	770	729	729	729
Coarse aggregate (20 mm)	663	663	663	627	627	627
Coarse aggregate (12.5 mm)	447	447	447	423	423	423
Water	160	160	160	162	162	162
Admixture	2.8	2.8	2.8	2.9	2.9	2.9
Microfiber	-	24.5	24.5	-	24.5	24.5
Macrofiber	-	-	49	-	-	49

The compressive strength, split tensile strength, modulus of rupture, modulus of elasticity, and rebar pull-out bond strength of the concrete were determined experimentally. The standard cubes of size 150 mm × 150 mm × 150 mm were used for the compression test. The split tensile strength and modulus of elasticity of concrete were determined using standard cylinder specimen of 150 mm diameter and 300 mm height. The prisms of size 100 mm × 100 mm × 500 mm were used for the modulus of rupture test. For rebar pull-out test, cylindrical specimens were cast with 10 mm bar embedded over a length of 100 mm. Three specimens were cast for each test and cured by immersing them in a water tank. The specimens were tested at the age of 28 days. The average of three specimens was calculated.

3. RESULTS AND DISCUSSIONS

A total of 90 specimens were cast and tested. The average value of compressive strength, split tensile strength, modulus of rupture, modulus of elasticity, and rebar pull-out bond strength of the concrete was calculated based on the test data of three specimens and is given in Table 2.

Table 2. Properties of recycled aggregate concrete.

Mix designation	Compressive strength (MPa)	Split tensile strength (MPa)	Modulus of rupture (MPa)	Modulus of elasticity (GPa)	Bond shear strength (MPa)
30-0-0	32.9	2.42	4.32	23.6	1.19
30-1-0	35.7	2.94	4.60	26.7	1.51
30-1-2	38.6	3.64	4.80	28.4	1.68
50-0-0	52.0	3.02	5.76	30.0	1.67
50-1-0	56.3	3.70	6.08	32.1	2.22
50-1-2	60.8	4.54	6.40	35.9	2.79

The compressive strength of recycled aggregate concrete was found to increase with the addition of steel fibers. The increase in compressive strength for 30-1-2 and 50-1-2 was found to 17.3% and 16.9%, respectively, when compared to the corresponding control mix having no fibers. An increase of 20% was found for modulus of elasticity in compression of both 30-1-2 and 50-1-2 when compared to the respective control mix having no fibers. The fibers will bridge across the micro-cracks and hence a confining stress will develop across the vertical cracks in the compression specimen. This helps to increase of vertical load resistance of compression specimens.

The increase in the split tensile strength and modulus of rupture was found to increase with increase in the fiber content. The increase in split tensile strength was found to 50% for both 30-1-2 and 50-1-2 when compared to the respective control mixes 30-0-0 and 50-0-0. Similarly, the increase in the modulus of elasticity was found to be 11% for both 30-1-2 and 50-1-2 when compared to the corresponding control mixes. The benefit of increase in the matrix strength in holding the fibers bridging across the crack could is not observed in higher strength grade (50 MPa). This may be due to the fact that the weak zones on the surface of recycled aggregate compensate the benefit of the increase in the pulling-out strength of steel fibers at the crack interface.

The addition of steel fibers to recycled aggregate concrete was found to be beneficial for increasing the bond strength with steel rebar. The increase in bond strength was found to 41% and 67% for 30-1-2 and 50-1-2, respectively, when compared to the corresponding control mix having no fibers. The resistance to rebar pull out in concrete is mobilized by intact concrete filled between the ribs of the rebar and the shear resistance at the rebar–concrete interface. The fibers hold resistance against the damage getting mobilized by the relative movement of rebar at the interface. This is the reason for having higher bond strength for fiber-reinforced concrete than the concrete having no fibers.

4. PREDICTION MODELS

The prediction models for compressive strength, split tensile strength, modulus of rupture, modulus of elasticity, and rebar pull-out bond strength were proposed based on the regression analysis of the test data and are given in Table 3. The dependent variables are characteristic strength of plain concrete (f_{ck}), percentage weight fraction of micro fibers (W_{fm}), and percentage weight fraction of macro fibers (W_{fM}).

The regression coefficient of W_{fm} and W_{fM} was found to be positive in all prediction models. This indicates that the addition of steel fibers is beneficial for increasing the various strength properties investigated in this study. This corroborates with the experimental data. The R^2 value of the prediction models was found to be 0.99 for all strength properties. Hence, it can be concluded that the proposed models account for 99% of the variation of the strength properties.

Table 3. Prediction models for the strength properties.

Property	Prediction model	R²
Compressive strength (MPa)	$f_{cu} = 1.057 f_{ck} + 3.736 W_{fm} + 1.850 W_{fM}$	0.99
Split tensile strength (MPa)	$f_{sp} = 0.453 \sqrt{f_{ck}} + 0.588 W_{fm} + 0.385 W_{fM}$	0.99
Modulus of rupture (MPa)	$f_r = 0.810 \sqrt{f_{ck}} + 0.259 W_{fm} + 0.130 W_{fM}$	0.99
Modulus of elasticity (GPa)	$E_c = 4.260 \sqrt{f_{ck}} + 2.668 W_{fm} + 1.375 W_{fM}$	0.99
Bond shear strength (MPa)	$\tau_{bd} = 0.036 f_{ck} + 0.419 W_{fm} + 0.185 W_{fM}$	0.99

5. CONCLUSIONS

Based on the experimental study on the strength properties of recycled aggregate concrete containing steel fibers, the following conclusions are arrived at:

- Addition of steel fibers is beneficial to improve the strength properties in all grades of concrete.
- The increase in strength properties in both 30 and 50 MPa grade concrete due to the addition of steel fibers is almost equal.
- Prediction based on the proposed models of strength properties accounts for the variation of test data.

REFERENCES

[1] Mi, R., Pan, G., Liew, K.M., and Kuang, T., "Utilizing recycled aggregate concrete in sustainable construction for a required compressive strength ratio," *J Clean Prod*, vol. 276, 2020. doi: 10.1016/j.jclepro.2020.124249.

[2] Thomas, J., Thaickavil, N. N., and Wilson, P. M., "Strength and durability of concrete containing recycled concrete aggregates," *J Build Eng*, vol. 19, pp. 349–365, 2018.

[3] Dimitriou, G., Savva, P., and Petrou, M. F., "Enhancing mechanical and durability properties of recycled aggregate concrete," *Const Build Mat*, vol. 158, pp. 228–235, 2018.

[4] Anikea, E. E., Saidanib, M., Olubanwob, A. O., Tyrera, M., and Ganjiana, E., "Effect of mix design methods on the mechanical properties of steel fibre reinforced concrete prepared with recycled aggregates from precast waste," *Structures*, vol. 27, pp. 664–672, 2020.

[5] Gao, D., Zhang, L., Zhao, J., and You, P., "Durability of steel fibre-reinforced recycled coarse aggregate concrete," *Const Build Mat*, vol. 232, 2020. doi: 10.1016/j.conbuildmat.2019.117119.

[6] IS:12262, *Ordinary Portland Cement 53 Grade – Specification*, New Delhi: Bureau of Indian Standards, 1987.

[7] IS:383, *Coarse and Fine Aggregate for Concrete – Specification*, New Delhi: Bureau of Indian Standards, 2016.

[8] IS:10262, *Concrete Mix Proportioning – Guidelines*, New Delhi: Bureau of Indian Standards, 2009.

36. Fracture Properties of Geopolymer Concrete Containing Recycled Aggregates

Cino Francis, Greeshma C and Job Thomas*
Department of Civil Engineering, School of Engineering, Cochin University of Science and Technology, Kochi, Kerala
*Corresponding author: job_thomas@cusat.ac.in

ABSTRACT: The fracture properties of geopolymer concrete containing recycled aggregates were studied. Fracture energy (G_f), critical Stress intensity factor (K_{Ic}), process zone length (l_p), and critical crack tip opening displacement ($CTOD_c$) were determined by three-point load testing of notched beam specimens. A total of 48 notched beam specimens were tested. The variables of the study were the grade of concrete and the addition of recycled aggregate. The fracture energy and critical stress intensity factor were found to increase with the grade of concrete and decrease with the addition of recycled aggregate. Similarly, the process zone length and CTOD were found to decrease with an increase in the grade of concrete and increase with the addition of recycled aggregates.

KEYWORDS: Geopolymer Concrete, Recycled Aggregate, Fracture Energy, Critical Stress Intensity Factor, Process Zone Length, Critical Crack Tip Opening Displacement

1. INTRODUCTION

Geopolymer concrete (GPC) is a promising alternative to conventional concrete. Geopolymer concrete is cement-free concrete and hence it is considered as green concrete. Recycled aggregate, which is obtained from demolished waste of concrete, is yet another green material used for the making of concrete. In this study, the geopolymer concrete containing recycled coarse aggregate (RCA) was prepared and tested to determine its fracture properties. Unlike the conventional strength properties, fracture properties are size independent. That means, the compressive strength of concrete will vary with the size of the specimen, but the fracture properties are evolved based on testing specimens with various sizes, and the size effect law is accounted for in the computation of the properties. In the case of conventional mechanical properties, the test results of laboratory-scaled specimens may not be directly applicable to the real-size specimens in the field as field sizes are quite larger when compared to the laboratory specimens. The effect of the size of specimens on the strength properties is given in Figure 1.

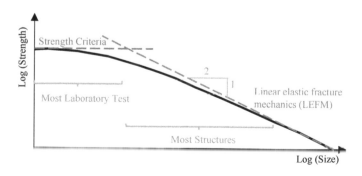

Figure 1. Size effect on the strength of concrete member.

Chapter 36 DOI- 10.1201/9781032657271-36

The fracture properties of two materials can be used for absolute or unbiased comparison of the material quality as it is independent of the size of the specimens tested. In this study, the fracture properties of geopolymer concrete containing recycled concrete aggregates are determined experimentally.

Choubey et al. (2016) found that there is a significant reduction in the fracture toughness of conventional concrete when recycled aggregates are added. Thomas and Sabu (2016) reported that the fracture properties of ambient cured geopolymer concrete are comparable with that of ordinary concrete of the same strength. Vishalakshi et al. (2018) observed that concrete made with granite aggregate having a rough surface exhibited higher fracture energy than limestone aggregate with smooth surface texture. The investigations of Peem et al. (2016) indicated that the inclusion of silica-rich materials had an adverse effect on the post-fire residual strength of geopolymer concretes made from recycled aggregates due primarily to the reduced porosity. Wang et al. (1998) showed that the initial curing under 80°C for 12–24 h is the optimum for geopolymer concrete containing recycled aggregate to achieve better strength properties. The investigations on the fracture properties of geopolymer concrete containing recycled aggregate are not found in the literature. The present study attempts to fill this gap. The fracture properties of the geopolymer concrete containing recycled aggregate were determined using notched beam specimens of different sizes.

2. METHODOLOGY

A total of 12 sets of notched specimens were tested varying the span to overall depth ratio and notch depth to overall depth ratio. Two numbers of span to overall depth ratios (l/d), namely, 6 and 8 were considered. Similarly, two numbers of notch depth to overall depth ratios ($a0/d$), namely, 0.2 and 0.4, were considered. A total of 38 notched beam specimens were prepared. The notched specimens are tested under three-point load. The test configuration of the specimens is shown in Figure 2.

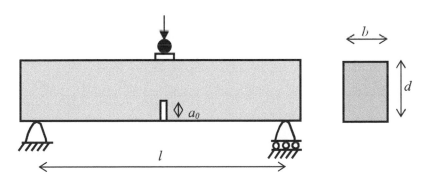

Figure 2. Test configuration of the notched beam.

The dimensional details of notched beam specimens are given in Table 1. Twelve specimens were tested and designated A1 to D3 as given in Table 1. The average failure load was used for the calculation of fracture properties. The fracture energy (Gf), critical stress intensity factor (KIc), fracture process zone length (lp), and critical crack tip opening displacement ($CTODc$) were calculated and reported. The companion cubes and cylinders of three each were also cast to test to find its compressive strength, split tensile strength, and modulus of elasticity of the concrete. Two concrete mixes were prepared, namely, control mix having natural crushed stone coarse aggregate alone and the second mix containing recycled aggregate alone. The recycled aggregate was made by crushing the old concrete of girders in a bridge project. The influence of replacing natural aggregate with recycled aggregate is studied.

Table 1. Dimensional details of the specimen.

Sample number	l/d ratio	$a0/d$ ratio	Span, l (mm)	Depth, d (mm)	Notch depth, $a0$ (mm)	Breadth, b (mm)	Overall length (mm)
A1	8	0.2	800	100	20	100	900
A2	8	0.2	1000	125	25	100	1100
A3	8	0.2	1200	150	30	100	1300
B1	6	0.2	600	100	20	100	700
B2	6	0.2	750	125	25	100	900
B3	6	0.2	900	150	30	100	1100
C1	8	0.4	800	100	40	100	900
C2	8	0.4	750	125	50	100	1100
C3	8	0.4	1200	150	60	100	1300
D1	6	0.4	600	100	40	100	700
D2	6	0.4	750	125	50	100	900
D3	6	0.4	900	150	60	100	1100

The mixes were designated to include the strength grade of concrete and the type of coarse aggregate used in it. The mix 30R represents 30 MPa concrete containing recycled (R) aggregate. The "N" in the designation indicates the mix containing natural aggregate. The details of the mix proportions are given in Table 2.

Table 2. Details of geopolymer concrete mix.

Constituents	Quantity of constituent materials in kg per cubic meter of concrete for the mix designated by			
	30N	30R	50N	50R
Geopolymer binder	510	510	600	600
Fine aggregate	650	650	590	590
Natural coarse aggregate (20 mm)	600	-	541	-
Natural coarse aggregate (12.5 mm)	400	-	360	-
Recycled coarse aggregate (20 mm)	-	560	-	500
Recycled coarse aggregate (12.5 mm)	-	370	-	335
Alkali activator	360	360	410	410

In this study, the geopolymer binder was prepared by mixing class F fly ash with ground granulated blast furnace slag (GGBS) in 1:1 proportion. The crushed stone fine aggregates of specific gravity 2.62 was used. The 20 and 12.5 mm nominal size coarse aggregates from granite quarry and having a specific gravity 2.69 and 2.71, respectively, were used as natural (N) aggregate in the control mix. Recycled (R) aggregates of nominal size 20 and 12.5 mm were found to have a specific gravity of 2.56 and 2.53, respectively. The mixture of sodium silicate ($Na2SiO3$) and sodium hydroxide (NaOH) solution was used as the alkali activator. The pH of the alkali activator was found to be 13.5 with a solid content of 45%.

3. CALCULATION OF FRACTURE PROPERTIES

The size-corrected strength (σ_N) of geometrically similar specimens proposed by Bazant and Planas (1990) is given by

$$\sigma_N = \beta f_t \left(1 + \frac{d}{D_0}\right)^{-1/2}$$
(1)

where   and D_0 are empirical coefficients D is the size of the beam. f_t is the tensile strength of concrete. The size effect law given by Eq. (1) has the advantage that it can be transformed into linear regression form (Bazant and Kazemi, 1990).

$$Y = AX + C$$
(2)

where A is the slope and C is the intercept of a linear fit between Y and every size X in Eq. (3). In which

$$Y = \left(f_t / \sigma_N\right)^2 ; X = d ; \beta = \sqrt{C} ; D_0 = C / A$$
(3)

The corresponding terms in Eqs. (2) and (3) can be equated. The magnitude of σ_N in Eq. (2) can be calculated by

$$\sigma_N = \frac{P_j}{b D_j}$$
(4)

where P_j is the failure load of the beam of width b and depth D_j. A and C can be obtained by regression analysis of the test data.

Fracture energy (G_f) is the energy required for the propagation of crack per unit area. The fracture energy (G_f) is calculated based on RILEM committee report (1990) given by

$$G_f = \frac{g'(\alpha) c_f}{\left(\beta f_t\right)^2 E} = C \frac{g'(\alpha) c_f}{E}$$
(5)

where c_f is half the length of fracture process zone and E is the modulus of elasticity of concrete. α is equal to a_0 / d and $g'(\alpha)$ is the derivative of non-dimensional energy release rate $g(\alpha)$. a_0 is the notch depth and d is the depth of the beam. The magnitude of c_f is calculated by

$$c_f = \frac{D_0 g(\alpha)}{g'(\alpha)} = \frac{C}{A} \frac{g(\alpha)}{g'(\alpha)}$$
(6)

and

$$g(\alpha) = \left(\frac{l}{d}\right) \pi \left[1.5 F(\alpha)\right]^2$$
(7)

where l is the span of the beam and d is the depth of beam. $F(\alpha)$ is the geometrical function and for three-point loading in notched beam specimen and is given by

$$when \frac{l}{d} = 4 , F(\alpha) = 1.090 - 1.735\alpha + 8.20\alpha^2 - 14.18\alpha^3 + 14.57\alpha^4$$
(8)

$$when \frac{l}{d} = 8 , F(\alpha) = 1.070 - 1.552\alpha + 7.71\alpha^2 - 13.55\alpha^3 + 14.25\alpha^4$$
(9)

The magnitude of $F(\alpha)$ corresponding to $\dfrac{l}{d} = 6$ is calculated by interpolation. The magnitudes of $g(\alpha)$ and $g'(\alpha)$ corresponding to test specimens are given in Table 3.

Table 3. Magnitudes of $g(\alpha)$ and $g'(\alpha)$.

$\alpha = a_0 / d$	$g(\alpha)$		$g'(\alpha)$	
	$\dfrac{l}{d} = 6$	$\dfrac{l}{d} = 8$	$\dfrac{l}{d} = 6$	$\dfrac{l}{d} = 8$
0.2	40.87	54.58	40.22	90.31
0.4	58.75	78.79	234.96	100.16

The stress intensity factor is the multiplier to account for the increase in magnitude of applied stress at the tip of the crack. The critical stress intensity is the value of the stress intensity required to propagate the crack at the tip. The critical stress intensity factor in Mode I fracture(K_{Ic}) is calculated based on plane stress condition and linear elastic fracture mechanics (LEFM) principles and is given by Eq. (10).

$$K_{Ic} = \sqrt{E_c G_f} \tag{10}$$

Crack tip opening displacement (CTOD) is the distance between the opposite faces of a crack tip at the 90° intercept position. The critical value of CTOD corresponds to the transition between ductile and brittle behavior and is denoted by $CTOD_c$. The magnitude of $CTOD_c$ is calculated by

$$CTOD_c = \sqrt{\frac{32 G_f c_f}{E \pi}} \tag{11}$$

The fracture process zone (FPZ) is defined as the region ahead of the traction-free crack tip. This region contains lots of distribute microcracks, in which the stress redistribution is quite complicated. The length of fracture process zone is calculated by

$$l_p = 2\, c_f \tag{12}$$

4. RESULTS AND DISCUSSIONS

The mechanical properties such as compressive strength, split tensile strength, and modulus of elasticity of geopolymer concrete are determined using standard specimens and the results. The average of three specimens is given in Table 4.

Table 4. The average mechanical properties of geopolymer concrete.

Mix designation	Compressive strength (MPa)	Split tensile strength (MPa)	Modulus of elasticity (GPa)
30 N	38.5	2.85	25.5
30 R	33.9	2.44	19.9
50 N	58.5	3.40	29.6
50 R	52.0	2.99	23.7

The compressive strength, split tensile strength, and modulus of elasticity of the geopolymer concrete containing recycled aggregate are found to be less than that of corresponding geopolymer concrete containing natural aggregate. The mortar sticking on the recycled aggregate will act as weaker zones in the mortar and aggregate interface. This is the reason for obtaining lower mechanical strength properties for geopolymer concrete containing recycled aggregate than that containing natural aggregate.

Fracture mechanics is used to characterize the loads on a crack, typically using a single parameter to describe the complete loading state at the crack tip. Fracture energy (G_f), critical stress intensity factor (K_{Ic}), process zone length (l_p), and critical crack tip opening displacement ($CTOD_c$) are the important fracture parameters of concrete. In this study, the fracture properties of the geopolymer concrete were found and are given in Table 5.

Table 5. Fracture properties of geopolymer concrete.

Mix designation	Fracture energy, G_f (J/m²)	Critical stress intensity factor, K_{Ic} (MPa√m)	Process zone length, l_p (mm)	Critical crack tip opening displacement, $CTOD_c$ (mm)
30 N	67.0	1.31	22.0	0.0172
30 R	58.6	1.08	26.6	0.0200
50 N	86.4	1.60	18.8	0.0167
50 R	76.2	1.34	21.9	0.0189

The variation in the fracture properties of geopolymer concrete was found to be similar to that of mechanical properties of concrete. The magnitude of fracture properties such as fracture energy and critical stress intensity factor increases with increase in grade of geopolymer concrete and decreases with the addition of recycled aggregate. The decrease in critical stress intensity factor was found to be 12.5% in 30 grade and 11.5% in 50 grade concrete when natural coarse aggregate was fully replaced with recycled aggregate. Similarly, the decrease in the critical stress intensity factor was found to be 17.5% for 30 grade concrete and 16.2% for 50 grade concrete with the replacement of natural aggregate with recycled aggregate. The length of fracture process zone and critical crack tip opening displacement decreases with the increase in grade of concrete and it increases with the addition of recycled aggregate to the concrete. The increase in the process zone length was found to be 20.9% in 30 grade concrete and 16.5% in 50 grade concrete with the addition of recycled aggregates. Similarly, the increase in the $CTOD_c$ was found to be 16.2% in 30 grade concrete and 13.1% in 50 grade concrete. In the process of crushing the old concrete for the preparation of recycled aggregate, mortar of old concrete will not get fully removed from the surface. This weak mortar will act as the path for the progress of fictitious crack. This may be the reason for lower values of fracture energy and critical stress intensity factor for concrete containing recycled aggregate. Similarly, the presence of weak zone in concrete aggregate surface makes the process zone larger and leads to larger crack tip opening displacement.

5. CONCLUSIONS

Based on the experimental study of the fracture properties of geopolymer concrete, following conclusions are arrived at.

- The fracture energy and critical stress intensity factor decreases with the addition of recycled aggregate.
- The fracture energy and critical stress intensity factor increases with grade of concrete.
- The process zone length and critical crack tip opening displacement increases with the addition of recycled aggregate.
- The process zone length and critical crack tip opening displacement decreases with increase in grade of concrete.

REFERENCES

[1] Bazant, Z. P., and Kazemi, M.,T., "Determination of fracture energy, process zone length and brittleness number from size effect, with application to rock and concrete," *Int J Frac*, vol. 44, pp. 111–131, 1990.

[2] Bazant, Z. P., and Planas, J., *Fracture and Size Effect in Concrete and Other Quasibrittle Materials*, New York: CRC Press, Routledge, 1998.

[3] Choubey, R. K., Kumar, S., and Rao, M., "Modelling of fracture parameters for crack, propagation in recycled aggregate concrete," *Const Build Mat*, vol. 106, pp. 168–178, 2016.

[4] Peem, N., Sata, V., and Chindaprasirt, P., "Influence of recycled aggregate on fly ash geopolymer concrete properties," *J Clean Prod*, vol. 112, pp. 2300–2307, 2016.

[5] RILEM Committee, "Size effect method of determining fracture energy and process zone size of concrete," *Mat Struc*, vol. 23, pp. 461–465, 1990.

[6] Thomas, J., and Sabu, N. J., "Fracture properties of OPC blended geopolymer concrete cured at ambient temperature: A review," Proceedings of the International Conference on Recent Advances in Civil Engineering (ICRACE '16), CUSAT, Kochi, pp. 467–471, 2016.

[7] Vishalakshi, K. P., Revathi, V., and Sivamurthy Reddy, S., "Effect of type of coarse aggregate on the strength properties and fracture energy of normal and high strength concrete," *Eng Frac Mech*, vol. 194, pp. 52–60, 2018.

37. Shear Strength of Cold-Bonded Aggregate Concrete Beams Containing Steel Fibres

Ajayakumar R, Abhirami Vikram, and Job Thomas*

Department of Civil Engineering, School of Engineering, Cochin University of Science and Technology, Kochi, Kerala

*Corresponding author: job_thomas@cusat.ac.in

ABSTRACT: The shear strength of cold-bonded aggregate concrete beam containing steel fibers was evaluated experimentally. A total of six beams were cast varying in the grade of concrete and quantity of steel fibers. Concrete grades of 30 and 40 MPa were used. Two types of steel fibers, namely, hooked-end macrofibers and straight microfibers, were used. The effect of addition on steel fibers on shear and compressive strength of concrete was determined. The test results showed that the increase in the shear strength is significant when steel fibers are added to the concrete. The prediction models were proposed for shear and compressive strength of concrete. The R-square value of the prediction model was found to be satisfactory.

KEYWORDS: Cold-Bonded Aggregate, Shear Strength, Compressive Strength, Steel Fibers

1. INTRODUCTION

The cold-bond aggregates are a substitute for the natural aggregates in concrete. This will help in protecting the environment by reducing the depletion of natural resources. The cold-bonded aggregate concrete is also brittle similar to that of conventional concrete, Addition of steel fibers can mitigate the brittle behavior of cold bonded aggregate. This study aims to find the effect of addition of steel fibers on strength characteristics of cold-bonded aggregate concrete. The shear strength of reinforced concrete beams containing cold-bonded aggregate and steel fibers was determined. Micro and macro steel fibers were used in this study.

Shear strength is an important property of concrete for the design of reinforced concrete structures. Al-Mahamoud et al. (2020) studied the shear behavior of reinforced concrete beams containing recycled coarse and fine aggregate. The decrease in shear strength was found to be 29% when natural fine and coarse aggregate in concrete is fully replaced with recycled fine and coarse aggregate. Pradhan et al. (2018) found that the overall crack pattern and failure mode of reinforced concrete beams containing recycled aggregate is similar to that of the beams containing natural aggregates. For the same longitudinal reinforcement ratio and shear span-to-depth ratio, the initial cracking load, ultimate failure load, and mid-span deflection were observed to be lower for recycled aggregate concrete beams when compared to the natural aggregate concrete beams. Uddin et al. (2019) evaluated the shear strength of reinforced concrete beam made with brick aggregate and recycled brick aggregate. The compressive and shear strength of concrete were found to decrease with the addition of brick aggregate. No significant difference in shear capacity of the reinforced concrete beams was found for beams made with brick aggregate and recycled brick aggregate. Etman et al. (2018) experimentally determined the shear strength of reinforced concrete beams containing recycled coarse aggregate. The decrease in shear capacity of beam was found to be 19% when 45% of natural aggregate in concrete is replaced with recycled aggregate. Rahal and Alrefaei (2018) found that the decrease in shear strength of reinforced concrete beam is 9% when the natural coarse aggregate is fully replaced with recycled aggregate.

Chapter 37 DOI- 10.1201/9781032657271-37

The studies on the influence of the addition of steel fibers on the shear strength of concrete beams containing cold-bonded aggregate are not been reported in the literature. There are no guidelines given in IS:456 (2000) for the design of shear of fiber-reinforced concrete. This study aims to determine the shear capacity of cold-bonded aggregate concrete beams containing steel fibers.

2. EXPERIMENTAL STUDY

A total of six concrete beams were cast and tested under two-point bending load setup. The cement, fine aggregate, cold-bonded coarse aggregate, and micro and macro fibers were used for making the concrete. Ordinary Portland cement of 53 grade having a specific gravity of 3.12 was used. Crushed-stone fine aggregate of specific gravity 2.70 was used. Cold-bonded coarse aggregates of nominal size of 20 mm were prepared with 80% fly ash and 20% cement. The dry mixture of fly ash and cement was rolled in a pelletizer and water was sprayed to it to get the fresh wet aggregate balls. These balls were then air-dried for 24 h and water cured for 7 days. The specific gravity of the cold-bonded aggregate was found to be 2.4. The cold-bonded aggregates are relatively round in shape. The picture of cold-bonded aggregate is given in Figure 1. Brass-coated straight microfibers having a length of 20 mm and a diameter of 0.22 mm were used. The hooked-end steel fibers of 30 mm length and 0.38 mm diameter were used as macro fibers. Superplasticizer containing sulfonated naphthalene polymers was used to get the required workability.

Figure 1. Cold-bonded fly ash aggregate.

Two grades of concrete, namely, M30 and M40 were prepared. The concrete mix was designed based on the absolute volume method given by IS:10262 (2019). The beam concrete mix was designated to indicate the grade, weight percentage of microfiber, and macrofiber. Thus, 30-1-2 indicates that concrete is of M30 grade containing 1% by weigh tof macro steel fibers and 2% by weight of micro steel fibers. The weight of the constituent materials per cubic meter of concrete is given in Table 1.

Table 1. Details of design mix of concrete.

Constituent materials	Weight of constituent materials in kg per cubic meter of concrete for the mix used in the beam					
	30-0-0	30-1-0	30-1-2	40-0-0	40-1-0	40-1-2
Admixture	2.85	2.85	2.85	3.1	3.1	3.1
Water	150	150	150	155	155	155
Cement	380	380	380	410	410	410
Fine aggregate	728	728	728	712	712	712

Cold-bonded coarse aggregate (20 mm)	1054	1054	1054	1035	1035	1035
Macrofiber (30 mm)	--	23.1	23.1	--	23.1	23.1
Microfiber (20 mm)	--	--	46.2	--		46.2

Corresponding to each mix, one beam and three companion cubes were cast and tested. The beams of 2200 mm long with cross-sectional dimension of 200 mm width and 300 mm depth were cast. The reinforcement cages were prepared and placed in an oiled steel mold. The constituent materials were mixed and the fresh concrete is poured into the molds and compacted using a needle vibrator. The beams and companion cubes were cured using moist burlap for 28 days and then tested. The reinforcement details of the beam and the loading arrangement are given in Figure 2. The test set-up of the beams is given in Figure 3. All beams were tested over a span of 2000 mm using 1000 kN beam testing machine. The companion cubes were also tested on the same day as that of the beam.

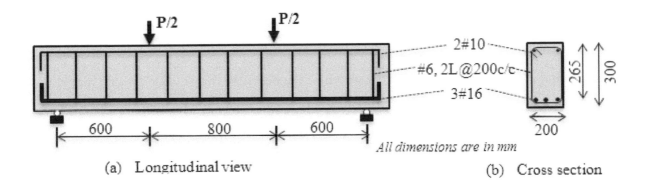

(a) Longitudinal view   (b) Cross section

Figure 2. Reinforcement details of test beams.

Figure 3. Test set-up of beams.

3. CALCULATION OF SHEAR STRESS

The failure load (P_u) of the beam was determined experimentally. The shear force (V_u) is determined using Eq. (1).

$$V_u = \frac{P_u}{2} \tag{1}$$

The shear resistance mobilized due to concrete (V_{uc}) is determined by Eq. (2).

$$V_{uc} = V_u - V_{us} \tag{2}$$

where is V_{us} the shear resistance mobilized due to stirrup and is determined by Eq. (3).

$$V_{us} = \frac{0.87 f_y A_{sv} d}{s_v} \tag{3}$$

where f_y is the yield strength of steel in stirrup, A_{sv} is the area of stirrup, d is the effective depth of beam and s_v is the spacing of stirrup. The shear strength mobilized in concrete (τ_c) is calculated using Eq. (4).

$$\tau_c = \frac{V_{uc}}{bd} \tag{4}$$

where b and d is the width and effective depth of the beam. In this study, b is 200 mm and d is 265 mm.

4. RESULTS AND DISCUSSIONS

A total of 6 beams and 18 companion cubes were cast and tested. The shear load on the beam was experimentally determined. The shear stress is calculated. The three companion cubes were tested on the same day of testing the beam and average cube compressive strength (f_{cu}) is calculated. The average value of compressive strength of concrete is determined. The shear force (V_u), shear resistance mobilized due to stirrup (V_{us}), shear resistance mobilized due to concrete (V_{uc}), shear strength mobilized in concrete (τ_c), and compressive strength of concrete(f_{cu}) are given in Table 2.

Table 2. Shear strength (τ_c) and compressive strength (f_{cu}) of concrete.

Beam mix designations	V_u (kN)	V_{us} (kN)	V_{uc} (kN)	τ_c (MPa)	f_{cu} (MPa)
30-0-0	90.0	27.0	63.0	1.19	31.8
30-1-0	102.5	27.0	75.5	1.42	32.8
30-1-2	105.0	27.0	78.0	1.47	35.1
40-0-0	97.5	27.0	70.5	1.33	42.5
40-1-0	112.5	27.0	85.5	1.61	43.6
40-1-2	115.0	27.0	88.0	1.66	47.9

The compressive strength of concrete (f_{cu}) is found to increase with addition of steel fibers. The increase in compressive strength in 30-1-2 is found to be 10% when compared to the control mix 30-0-0 having no fibers. This may be due to the fact that the fibers bridging across the crack mobilize the resistance against the lateral expansion, which helps to induce confinement in the concrete. The strength of confined concrete will be greater than unconfined concrete. This is the reason for getting higher strength for fiber-reinforced concrete when compared to concrete having no fibers.

The shear strength is also found to increase with the addition of steel fibers. The increase in shear strength of beam 30-1-2 is found to be 23.5% when compared to beam 30-0-0 and similarly, the increase in the shear strength of beam 40-1-2 is found to be 24.8%. The bridging of the fiber across the cracks mobilizes shear resistance in the concrete. The bond between the steel fibers and the concrete increases with increase in compressive strength. This is the reason for higher percentage increase of shear resistance in higher grades of concrete.

The crack pattern in the beams at ultimate stage is given in Figure 4. In the initial stages of loading, cracks formed in the middle of the span. In subsequent stages of loading, the diagonal cracks formed in the shear span of the beams on either side. One of the diagonal cracks opened up and lead to the failure of the beam. All the beams failed in diagonal shear. More number of cracks were found in fiber-reinforced concrete beams when compared to the corresponding beams without any fiber. This indicates that the fibers are effective in redistributing the stress in the concrete.

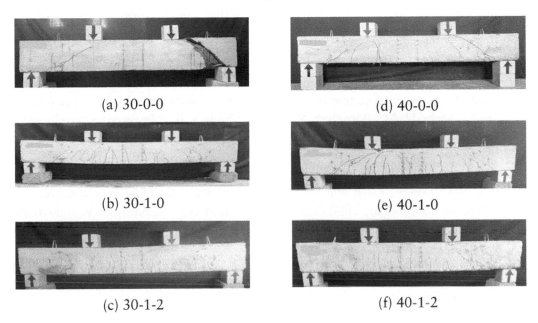

(a) 30-0-0 (d) 40-0-0

(b) 30-1-0 (e) 40-1-0

(c) 30-1-2 (f) 40-1-2

Figure 4. *Crack pattern in test beams at ultimate stage.*

5. PREDICTION OF PROPERTIES

The regression analysis is carried out to develop the compressive strength and shear strength of cold-bonded concrete containing steel fibers. The prediction models for strength properties of fiber-reinforced concrete are given in Table 3.

Table 3. Prediction model for strength properties.

Strength property	Prediction model*	R^2
Compressive strength (MPa)	$f_{cuf} = f_{cu0}\left(1 + 0.029 W_{fM} + 0.043 W_{fm}\right)$	0.980
Shear strength (MPa)	$\tau_{cf} = 0.197\sqrt{f_{cu0}} + 0.259 W_{fM} + 0.024 W_{fm} + 0.065$	0.995

The compressive strength of fiber-reinforced concrete (f_{cuf}) can be predicted if parameters such as cube compressive strength of plain concrete (f_{cu0}), weight fraction of macrofiber in percentage (W_{fM}), and weight fraction of microfibers in percentage (W_{fm}) are known. In other words, the quantity of fibers needs to be added can be calculated to achieve a desired strength (f_{cuf}) for a given grade of concrete. The prediction model for shear strength of fiber-reinforced concrete (τ_{cf}) is given in Table 3. The R-square value of the prediction models given in Table 3 is found to be greater than 98%. Hence it can be concluded that the proposed models are good in predicting the strength properties of cold-bonded aggregate containing steel fibers.

6. CONCLUSION

Based on the experimental study on the cold-bonded aggregate concrete containing steel fibers, following conclusions are arrived at

- The addition of steel fibers is beneficial to improve the compressive strength of concrete.
- The increase in shear strength is significant when steel fibers are added.
- The proposed prediction models can reasonably reflect the variations in the experimental data.

REFERENCES

[1] Al-Mahmoud, F., Boissiere, R., Mercier, C., and Khelil, A., "Shear behaviour of reinforced concrete beams made from recycled coarse and fine aggregates," *Structures*, vol. 25, pp. 660–669, 2020.

[2] Etman, E. E., Afefy, H. M., Baraghith, A. T., and Khedr, S. A., "Improving the shear performance of reinforced concrete beams made of recycled coarse aggregate," *Const Build Mat*, vol. 185, pp. 310–324, 2018.

[3] IS:10262, *Concrete Mix Proportioning – Guidelines*, New Delhi, India: Bureau of Indian Standards, 2019.IS:456, *Plain and Reinforced Concrete – Code of Practice*, New Delhi, India: Bureau of Indian Standards, 2000.Pradhan, S., Kumar, S., and Barai, S. V., "Shear performance of recycled aggregate concrete beams: An insight for design aspects," *Const Build Mat*, vol. 178, pp. 593–611, 2018.

[4] Rahal, K. N., and Alrefaei, Y. T., "Shear strength of recycled aggregate concrete beams containing stirrups," *Const Build Mat*, vol. 191, pp. 866–876, 2018.

[5] Uddin, M. T., Shikdar, M. K. H., and Joy, J. A., "Shear strength of RC beams made with brick aggregate without shear reinforcement," *J Civil Eng*, vol. 47, pp. 31–45, 2019.

38. Bond Strength Between Brick and Mortar

Gary Thomas Job[*]

Department of Civil Engineering, Indian Institute of Technology Madras, Chennai, Tamil Nadu, India

[*]Corresponding author: garythomasjob@gmail.com

ABSTRACT: Behavior of masonry greatly depends on the characteristics of masonry units, mortar, and the bond between them. The bond strength between the masonry unit and mortar depends on the mortar strength and the surface characteristics and water absorption of the masonry unit. In this study, the bond strength of the masonry made up of two types of bricks and three types of mortar is determined experimentally. The burned clay wire-cut bricks and hand-molded bricks were used in this study. The triplet specimen was used to determine the bond strength of the masonry specimen. The bond strength on days 7 and 28 was determined. The bond strength was found to increase with an increase in mortar strength in both wire-cut and hand-molded brick masonry specimens. Based on the study, a regression analysis model for predicting the bond shear strength of burned clay brick specimens is proposed.

KEYWORDS: Brick, Masonry, Bond Strength, Hand-Molded Bricks, Wire-Cut Bricks

1. INTRODUCTION

Construction using brick masonry is very common all over the world. Codes of practice of masonry design give lines to assess masonry compressive strength by considering compressive strength, height, and mortar type of masonry units. It is assumed that there is a perfect bond between the masonry unit and the mortar. The masonry failure in earthquakes indicates that the weakest link in the masonry assemblage is the interface between mortar and masonry units (Nayak and Dutta, 2016). Hence, importance is identified to study the bond between masonry unit and mortar.

2. REVIEW OF LITERATURE

The various locally available materials are being used for the construction of masonry elements in the building. A lot of research works are reported to establish the properties of the masonry by changing the characteristics of constituent materials. Recent literature on the bond behavior of masonry is reviewed. Murthi et al. (2022) determined the shear bond strength of masonry using triplet specimens. The optimum replacement of sand with fine recycled aggregate was determined when subjected to axial compression. Jiao et al. (2019) found that the bond and compressive strength of the masonry is directly related. Raj et al. (2020) reported that the weak mortar and cement slurry coating on autoclaved aerated concrete blocks gives better triplet bond strength. Singh and Munjal (2017) observed that the bond strength of concrete masonry is relatively low due to the less contact area. Hamdy et al. (2019) found that the addition of silica fume or polypropylene fibers enhances the bond strength of masonry even in adverse exposure conditions. The experimental works of Barr et al. (2018) indicate that the bond strength of masonry is directly proportional to the hydraulicity of the binder and the pre-wetting time of the masonry unit. Sarhosis and Sheng (2014) developed two-dimensional computational models for the masonry based on large experimental data in order to predict material parameters for low-bond strength masonry. Sarhosis et al. (2015) demonstrated that the mechanical behavior of masonry brickwork lintel is greatly influenced by the bond between brick mortar interface. Luso and Lourenco (2017) observed that the average tensile bond strength of masonry is 0.8 MPa when grouting is done

Chapter 38 DOI- 10.1201/9781032657271-38

in the masonry. Soleymani et al. (2022) found that brick–mortar bond strength is highly influenced by mortar characteristics. Murthi et al. (2021) reported that the bond strength failure is exhibited with the increase in the thickness of mortar in masonry. The review of the literature indicates that the strength of mortar and surface characteristics of brick is greatly influenced by the interface bond strength of the masonry. In this paper, the influence of the mortar strength and mortar thickness on the bond strength of brick–mortar interface is studied.

3. MATERIALS AND METHOD

The OPC 53 grade cement, recycled finer sand, and burned clay bricks were used for the experiment. The specific gravity and 28th day compressive strength of cement was found to be 3.05 and 57 MPa, respectively. Crushed stone sand conforming to Zone II of IS 383 (2016) was used. The specific gravity of the sand was found to be 2.70. The bulk density of the fine aggregate is found to be 1550 kg/m³.

The bond strength of the masonry was determined using triplet specimen. The brick masonry prism was prepared with varying joint mortar. The two bricks were used, namely, wire-cut brick and hand-molded brick. Both types of bricks were burned in a kiln. These bricks are purchased from local market. The size of both wire-cut and hand-molded brick was 210 mm × 98 mm × 65 mm. Hand-molded brick had a frog on the top. The average compressive strength and water absorption of wire-cut bricks were found to be 9.5 MPa and 14% and that of hand-molded bricks are 4.5 MPa and 19%. Three mortar mix proportions, namely, M1, M2, and M3 were used. The cement-to-fine aggregate ratio of mix M1 was 1:4 and M2 was 1:6, and M3 was 1:8 by weight. The mixing water quantity was taken as 0.4 by the weight of mortar. The 7th day average compressive strength of the mortar M1, M2, and M3 was found to be 7.5, 5.0, and 3.0 MPa, respectively. Similarly, the 28th day average compressive strength of mortar was found to 10.5, 7.0, and 4.3 MPa for M1, M2, and M3 mortar mixes. The details of the triplet specimens are given in Table 1 and three triple specimens were prepared for each set. A total of 36 triplet specimens were cast. The bricks are immersed in water for 24 hours and placed in ambient condition for few minutes to make it surface dry condition. The triplet specimens are fabricated with these saturated surface dry bricks. A mortar of 15 mm thickness was provided. The specimens are cured from the next day of casting by sprinkling the water three to four times a day. The specimens are tested at the age of 7 or 28 days. The test configuration is given in Figure 1.

Table 1. Designation and details of the specimen.

Sl. no.	Specimen designation	Brick type[a]	Mortar type	Age of testing (days)
1	W-M1-7	W	M1	7
2	W-M1-28	W	M1	28
3	W-M2-7	W	M2	7
4	W-M2-28	W	M2	28
5	W-M3-7	W	M3	7
6	W-M3-28	W	M3	28
7	H-M1-7	H	M1	7
8	H-M1-28	H	M1	28
9	H-M2-7	H	M2	7
10	H-M2-28	H	M2	28
11	H-M3-7	H	M3	7
12	H-M3-28	H	M3	28

[a]W = wire-cut bricks; H = Hand-molded brick.

The specimens were rotated at 90° from the cast direction in the triplet test. The failure load is noted and the bond strength of the specimen is calculated using Eq. (1). The average of three specimens was calculated using Eq. (1).

Bond strength (f_b) = Failure load (P)/shear plane area (A_s) (1)

Figure 1. *Testing of triplet specimen W-M1-28.*

4. RESULTS AND DISCUSSIONS

A total of 36 specimens were tested. The average bond strength of three specimens was calculated and reported. The bond strength of the specimen and the average strength and standard deviation of the test results are given in Table 2.

Table 2. Bond strength between masonry unit and bed joint mortar.

Specimen designation	Bond strength in MPa				Standard deviation (MPa)
	Specimen 1	Specimen 2	Specimen 3	Average strength	
W-M1-7	0.0826	0.0850	0.0875	0.08503	0.0024
W-M1-28	0.0972	0.1020	0.1045	0.10123	0.0037
W-M2-7	0.0656	0.0705	0.0705	0.06883	0.0028
W-M2-28	0.0777	0.0826	0.0850	0.08179	0.0037
W-M3-7	0.0437	0.0486	0.0510	0.04778	0.0037
W-M3-28	0.0510	0.0559	0.0583	0.05507	0.0037
H-M1-7	0.0875	0.0899	0.0923	0.08989	0.0024
H-M1-28	0.1045	0.1093	0.1118	0.10852	0.0037

H-M2-7	0.0729	0.0777	0.0802	0.07693	0.0037
H-M2-28	0.0875	0.0948	0.0972	0.09313	0.0050
H-M3-7	0.0534	0.0559	0.0607	0.05668	0.0037
H-M3-28	0.0583	0.0607	0.0705	0.06316	0.0064

4.1. Failure of specimen

All the specimens failed in double shear. That means, there were two shear planes in the specimens and is given in Figure 2. This indicated that the distribution of the loads to the support in the specimen was symmetrical. The cracks were found close to the plane face of the brick and the brick. In hand-molded bricks, the bond strength offered by the mortar key into the frog of hand-molded brick was found to give better bond strength than the plane surface, this was indicated by the formation of crack at the plane surface of brick and mortar.

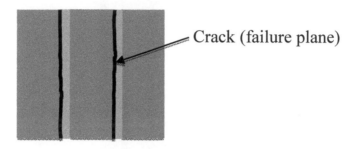

Crack (failure plane)

Figure 2. Cracking in the triplet specimen W-M1-28.

4.2. Effect of mortar strength

The bond shear strength of triplet specimens is also found to increasing with the increase of mortar strength. The magnitude of bond strength and standard deviation of the test data are given in Table 2. The variation of triplet bond strength of masonry for various mortar strength is shown in Figure 3. The average increase in bond strength of mix M1 (1:4) was to be 80% on the 7th day and 67% on the 28th day when compared to the strength of mix M3 (1:8). This is due to the fact that the adhesion of mortar to the brick increases with the increase in cement content in the brick.

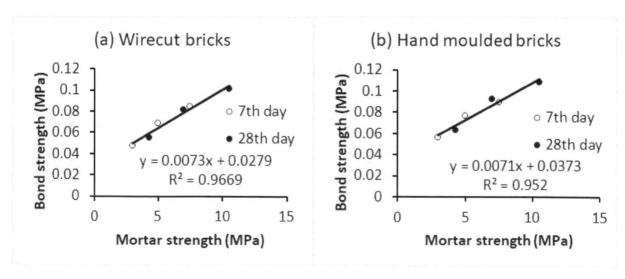

Figure 3. Effect of mortar strength on the bond strength of masonry.

5. PREDICTION MODELS

The prediction model for the bond strength of masonry (f_{bm}) was proposed based on the linear regression analysis and is shown in Figure 3.

For wire-cut bricks,

$$f_{bm} = 0.0073\, f_m + 0.0279 \tag{2}$$

For hand-molded bricks,

$$f_{bm} = 0.0071 f_m + 0.0373 \tag{3}$$

where f_m is the mortar strength. The statistical R-square value of regression is found to be above 95% for both Eqs. (2) and (3). The difference in the magnitude of the coefficient of the regression coefficient in the prediction Eqs. (2) and (3) is found to be small. This may be due to the fact that the surface characteristics of the wire-cut and molded brick differ slightly.

6. CONCLUSIONS

The following conclusions are arrived at based on the experimental study conducted:

- The bond strength of the masonry increases with an increase in the mortar strength.
- The bond strength of masonry made up of wire-cut and the hand-molded clay brick is almost the same.

In the future, the experiment can be conducted to evaluate the bond strength of cement block masonry prisms, for which the surface characteristics of the masonry unit are significantly different from that of clay bricks.

REFERENCES

[1] Barr, S., McCarter, W. J., and Suryanto, B., "Bond-strength performance of hydraulic lime and natural cement mortared sandstone masonry," *Const Build Mat*, vol. 84, pp. 128–135, 2018.

[2] Hamdy, G. A., El-Hariri, M. O. R., and Farag, M. F., "Use of additives in mortar to enhance the compression and bond strength of masonry exposed to different environmental conditions," *J Build Eng*, vol. 25, paper ID: 100765, 2019.

[3] IS:383, *Coarse and Fine Aggregate for Concrete-Specification*, New Delhi: Bureau of Indian Standards, 2016.

[4] Jiao, Z., Wang, Y., Zheng, W., Huang, W., and Zhao, Y., "Bond properties of alkali-activated slag concrete hollow block masonry with different mortar strength grades," *Const Build Mat*, vol. 216, pp. 149–165, 2019.

[5] Luso, E., and Lourenco, P. B., "Bond strength characterization of commercially available grouts for masonry," *Const Build Mat*, vol. 144, pp. 317–326, 2017.

[6] Murthi, P., Krishnamoorthi, S., Poongodi, K., and Saravan R., "Development of green masonry mortar using fine recycled aggregate based on the shear bond strength of brick masonry," *Mat Today Proc*, vol. 61, no. 2, pp. 413–419, 2022.

[7] Murthi, P., Md. Akib, Md. Imran, Ahmed, S., and Prasanna, V., "Studies on the strength variation of brick masonry using novel blended masonry mortar mixes and mortar thickness," *Mat Today Proc*, vol. 39, pp.126–130, 2021.

[8] Nayak, S., and Dutta, S. C., "Failure of masonry structures in earthquake: A few simple cost effective techniques as possible solutions," *Eng Struct*, vol. 106, pp. 53–67, 2016.

[9]   Raj, A., ChBorsaikia, A., and Dixit, U. S., "Bond strength of autoclaved aerated concrete (AAC) masonry using various joint materials," *J Build Eng*, vol. 28, paper ID:101039, 2020.

[10]   Sarhosis, V., and Sheng, Y., "Identification of material parameters for low bond strength masonry," Eng Struct, vol. 60, pp. 100–110, 2014.

[11]   Sarhosis, V., Garrity, S. W., and Sheng, Y., "Influence of brick–mortar interface on the mechanical behaviour of low bond strength masonry brickwork lintels," *Eng Struct*, vol. 88, pp. 1–11, 2015.

[12]   Singh, S. B., and Munjal, P., "Bond strength and compressive stress-strain characteristics of brick masonry," *J Build Eng*, vol. 9, pp.10–16, 2017.

[13]   Soleymani, A., Najafgholipour, M. A., and Johari, A., "An experimental study on the mechanical properties of solid clay brick masonry with traditional mortars," *J Build Eng*, vol. 58, Paper ID: 105057, 2022.

39. Study on the Effect of Water–Cement Ratio on Accelerated Concrete Curing

Subha Vishnudas[1,*], Neenu Maria Jose[1], and Ushakumari G[2]

[1]Faculty of Civil Engineering, School of Engineering, Cochint University of Science and Technology, Cochin, Kerala

[2]PWD, Kerala

*Corresponding author: v.subha@cusat.ac.in

ABSTRACT: The compressive strength of cement concrete obtained after 28 days of moist curing is used in the quality control of constructions. However, for economical quality control, for reworking before the concrete gets hardened, and for reducing the waiting time, finding the 28-day compressive strength at an earlier time with a reasonable accuracy is necessary. The 28-day compressive strength of concrete can be estimated early by finding the strength at an early age with accelerated curing. For the prediction of 28-day compressive strength, a previously established correlation curve between the early strength with accelerated curing and the 28-day strength standard method is suggested by IS: 9013-1978. Major research world has proved that a substantial portion of compressive strength is attained due to accelerated curing and accelerated curing techniques can be used for early prediction of compressive strength of concrete. This paper shows the relation between accelerated curing strength and normal curing strength and the influence of water–cement ratio on 53-grade concrete.

KEYWORDS: Compressive Strength, Accelerated Curing, Accelerated Curing Strength, Normal Curing Strength, Predicted Strength

1. INTRODUCTION

Many different types of cement are available in the market, and design mixes currently have significant advantages over nominal mixes. In India, it is frequently difficult to obtain materials from the same source throughout the construction process. A new mix design has to be adopted when a material's source changes, which slows down execution. Concrete's quality in construction projects is assessed by its 28-day compressive strength, for this test specimens need to be moist-cured for 28 days prior to testing. For many years, a substantial research initiative has been focused on early predicting the 28-day compressive strength of concrete. It has been found that a reliable and efficient method is required for assessing concrete in the field using accelerated curing procedures (Krishna Rao et al., 2010). By measuring the strength at an early age with accelerated curing, the 28-day compressive strength of concrete can be predicted (Neelakantan et al., 2013).

In the prefabrication sector, where high early strength is attained and allows the removal of formwork within 24 h, accelerated curing procedures are very beneficial. This decreases the cycle time and has a positive impact on cost savings. Physical procedures and the use of admixtures as catalysts for the hydration process are two separate ways to speed up the curing process. Combinations of the following physical methods are typically employed to speed up the curing process: (a) raising the curing temperature and (b) adding moisture to the curing environment (Weider, 2004). There are many techniques, which include steam curing, electrical resistance heating, and conductive/convective heating (low and high pressure). The use of elevated curing temperatures still continues to be the main technique

Chapter 39 DOI- 10.1201/9781032657271-39

for accelerated curing in the precast industry today. The pace of hydration is significantly influenced by the concrete's curing temperature (Liu, 2012). Higher early strength and decreased long-term strength are correlated, as per a study conducted by Vollenweider et al. (2004). High curing temperatures have a favorable effect on early mechanical property development but they have a negative impact on strength later on. Higher temperatures cause a rapid initial rate of hydration, which could delay further hydration. For design purposes, high early strength, particularly at 1 day, was taken into account and used to anticipate 28-day curing strength (Gesoglu et al., 2013).

This study examines the effectiveness of accelerated curing procedures in predicting 28-day concrete strength at an early age and the relationship between accelerated curing strength and normal curing strength on OPC 53 grade (IS-12269) for three water–cement ratios of 0.55, 0.50, and 0.45.

2. MATERIALS AND METHODS

2.1. Materials

The materials used in this study are cement, fine aggregate, and coarse aggregate. All the materials used were tested as per respective IS codes to assess their engineering properties and the results were compared with those in the relevant IS code. The following IS codes were used in this study: IS 9013:1978 (2004), IS 12269:2013 (2013), IS 4031 (part 4):1988 (2005), IS 1489 (part 1): 1991 (2005), IS 2386:1963 (part 3) (2005), IS 2386:1963 (part 4) (2005), IS 3495:1992 (part 1) (2005), IS 383:1970 (2002), IS 456:2000 (2005), and IS 516:1959 (2005).

2.1.1. Material characterization

2.1.1.1. Cement

For analyzing the properties of Ordinary Portland Cement, OPC 53 grade following tests were conducted: consistency, fineness, setting time, specific gravity, and compressive strength. The properties of OPC 53 grade are presented in Table 1.

Table 1. Properties of OPC 53 grade cement.

Sl. no:	Property	OPC 53 cement	IS limits
1	Fineness	5%	<10%, IS:1489-Part-I
2	Standard consistency	29.72%	IS:4031(Part4)-1988
3	Initial setting time (minutes)	225	IS:1489-Part-I >30 min
4	Final setting time (minutes)	440	IS:1489-Part-I <600 min
5	Specific gravity	2.95	IS:4031:1988 3–3.15
6	Compressive strength at 7 days (N/mm^2)	37.20	IS:4031:1988

2.1.1.2. Fine aggregates and coarse aggregates

The fine aggregate used was M sand. Tests were conducted as per IS: 2386-1970 part-III and the properties are found to be as follows specific gravity (2.75), bulk density (1602 kg/m^3), porosity (0.382), and void ratio (0.618). Gradation and testing of the fine aggregate have been done as per IS: 2386-1963. Figure 1 shows the particle size distribution graph of fine aggregates. For coarse aggregate, the properties are specific gravity (3.12), bulk density (1780 kg/m^3), porosity (0.433), void ratio (0.762), crushing value (23.69%), abrasion value (25.4%), and impact value (16.31%). Gradation and testing of the coarse aggregate have been done as per IS: 2386-1963. Figure 2 shows the particle size distribution graph of coarse aggregates.

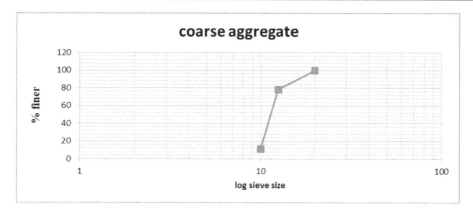

Figure 1. Particle size distribution of fine aggregate (M sand).

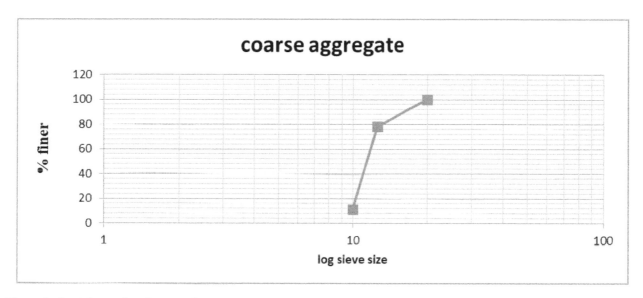

Figure 2. Particle size distribution of coarse aggregate.

2.2. Mix design

Various trial mixes were conducted for the selection of proportion of water and cement (Tables 2 and 3, IS: 456(2000)), coarse aggregate and fine aggregate (refer Clause 4.4, A-7 and B-7 and Table 2 of IS: 383(1970)). Ordinary Portland cement OPC 53 grade (IS-12269 (2013)) for three water–cement ratios 0.55, 0.50, and 0.45 were considered. Final mix corresponding to acceptable slump test result based on the workability and durability criteria of IS: 456 (2000). The obtained slump value for OPC 53 grade are 7.6 mm for 0.45 w/c ratio, 8.2 mm for 0.5 w/c ratio, and 9.0 mm for 0.55 w/c ratio. Table 2 shows details of the mix design.

Table 2. Details of mix design.

Sl. no.	Cement type	W/C ratio	Water (l)	Fine aggregate (kg)	Coarse aggregate (kg)	Cement (kg)	Mix proportion	No. of specimen
1	OPC 53	0.45	200	676.04	1274.60	444.00	1: 1.52: 2.87	6
2	OPC 53	0.50	200	690.51	1301.90	400.00	1: 1.73: 3.25	6
3	OPC 53	0.55	200	702.36	1324.20	364.00	1: 1.93: 3.64	6

2.3. Experimental setup

2.3.1. Curing methods

The prediction of compressive strength at an early age, cube specimens of the same sample were tested after two methods of curing.

(a) Accelerated curing: Boiling water method as per the IS: 9013 (1978)

As per the IS: 9013 (1978), the procedure is: after the specimens are cast (150 mm cubes), they should be kept in a humid environment (90% humidity and 27+2°C for 23 h + 15 min). The specimens shall then be gently lowered into the curing tank containing boiling water and shall be kept in that condition for 3½ h + 5 min. Then the specimens shall be removed from the boiling water, removed from the mold after 24 h and cooled by immersing in water (27+2°C) for 2 h. The specimens shall then be tested for compressive strength. The age of the cubes at the time of testing shall be 28½ h + 20 min.

(b) Conventional curing

Normal curing for 28 days by immersing in cold water at room temperature, after removing from the mold after 24 h as per IS 456 (2000).

2.3.2. Test for compressive strength

The test was conducted in a compression testing machine of sufficient capacity and capable of applying the load at the rate specified in IS: 516-1959. Six cubes were cast for each w/c ratio and among those three cubes for testing accelerated curing strength after boiling water curing and three cubes for testing for testing normal curing strength after conventional curing.

3. RESULTS AND DISCUSSIONS

The results of the test conducted in a compression testing machine of sufficient capacity have been tabulated in Table 3, which includes accelerated curing strength, predicted strength, and 28 day normal curing strength. The equation for calculating the predicted strength after measuring accelerated curing strength is shown in Eq. (1) (IS: 9013-1978).

$$R_{28}=1.64*R_a+8.09 \tag{1}$$

where R_a is the accelerated curing strength, R_{28} is the predicted strength.

Table 3. Compressive strength.

Sl. no	Type of cement	WC ratio	Acc. curing strength at @28 h (N/mm²)	Predicted strength (N/mm²)	28 normal curing strength (N/mm²)	Strength attained due to acc. curing (%)
1	OPC 53 Grade	0.55	12.36	28.36	31.04	39.82
2	OPC 53 Grade	0.55	12.71	28.93	31.8	39.97
3	OPC 53 Grade	0.55	12.53	28.63	32.96	38.07
4	OPC 53 Grade	0.5	15.09	32.84	39.24	38.45
5	OPC 53 Grade	0.5	15.76	33.94	38.27	41.18
6	OPC 53 Grade	0.5	16.2	34.69	38.62	41.95
7	OPC 53 Grade	0.45	19.05	39.34	40.89	46.59
8	OPC 53 Grade	0.45	20.01	40.9	41.2	48.57
9	OPC 53 Grade	0.45	19.8	40.56	42.27	46.85

Figure 3 shows the graphical representation of values of accelerated strength, predicted strength, and normal 28 day curing strength for various mix proportions using Ordinary Portland Cement of Grade 53. Normal curing strength is higher than the predicted strength for all the different mix proportions considered for the study. From the figure, it can also be seen that a specified portion of the strength to be attained at 28 day is achieved by accelerated curing process within 28 h which justify the findings of Gesoglu (2013). A substantial portion of the strength is achieved due to accelerated curing techniques.

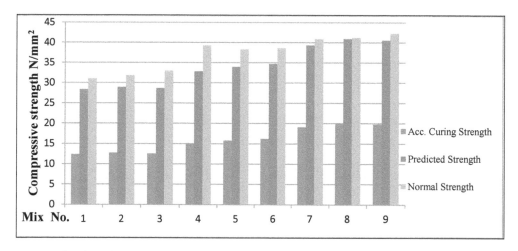

Figure 3. Experimental and predicted values of compressive strength for OPC 53 grades of concrete.

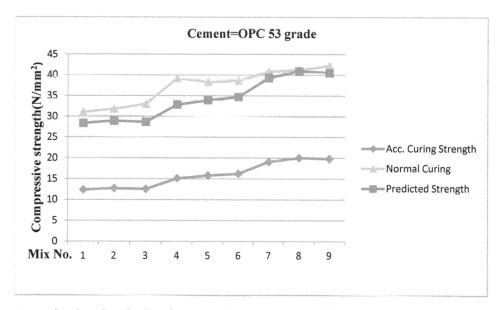

Figure 4. Experimental and predicted values for cement OPC 53 grade at different w/c ratio.

Experimental and predicted values for cement OPC 53 grade at different w/c ratios are shown in Figure 4. In this case, the variation is not at all a linear one. From the results obtained in the experimental study, the corresponding plot shows that a substantial portion of the 28 day compressive strength is attained in short period due to accelerated curing. In this case, this type of cement strength attained is higher than the predicted strength. In the case of 7, 8, and 9 mixes, the difference between predicted strength and normal curing strength is negligible. But it is observed that a correlation does exist between the compressive strengths at 28 h after accelerated curing and at 28 day after normal curing, but a generalized linear equation may not help in a realistic prediction.

4. CONCLUSIONS

From the experimental investigation following conclusions were drawn,

i. Actual compressive strength after conventional curing is greater than the strength predicted(accelerated curing) using the correlation equation as per IS: 9013.

ii. Strength varies at different proportions in different water–cement ratio. In accelerated curing method and conventional curing as water–cement ratio increases strength decreases. Compressive strength at water–cement ratio 0.45 is higher than 0.50 and 0.55.

iii. Strength varies at different proportions in concrete with different types of w/c ratio. In the case of 7, 8, and 9 mixes, the difference between predicted strength and normal curing strength is negligible.

REFERENCES

[1] Brent vollen Weider, "Various methods of accelerated curing for concrete applications and their impact on short term and long term compressive strength," *CE 241: Concrete Technology*, Paper-1, 2004.

[2] IS 12269:2013. *Ordinary Portland Cement, 53 Grade—Specification*. New Delhi: Bureau of Indian Standards, 2013.

[3] IS 1489 (part 1): 1991 (reaffirmed in 2005). *Portland Pozzolana Cement—Specification*. New Delhi: Bureau of Indian Standards, 2005.

[4] IS 2386:1963 (part 3). *Methods of Test for Aggregate for Concrete, Specific Gravity, Density, Voids, Adsorption and Bulking*. New Delhi: Bureau of Indian Standards, 2005.

[5] IS 2386:1963 (part 4). *Methods of Test for Aggregate for Concrete, Mechanical Properties*. New Delhi: Bureau of Indian Standards, 2005.

[6] IS 3495:1992 (part 1). *Methods of Test of Burnt Clay Building Bricks, Determination of Compressive Strength*. New Delhi: Bureau of Indian Standards, 2005.

[7] IS 383:1970 (reaffirmed in 2002). *Specification for Coarse and Fine Aggregates From Natural Sources for Concrete*. New Delhi: Bureau of Indian Standards, 2002.

[8] IS 4031 (part 4):1988 (reaffirmed in 2005). *Method of Physical Tests for Hydraulic Cement*. New Delhi: Bureau of Indian Standards, 2005.

[9] IS 456:2000 (reaffirmed in 2005). *Plain and Reinforced Concrete—Code of Practice*. New Delhi: Bureau of Indian Standards, 2005.

[10] IS 516:1959. *Methods of Test for Strength of Concrete*, New Delhi: Bureau of Indian Standards, 2005.

[11] IS 9013:1978 (reaffirmed in 2004). *Method of Making, Curing and Determining the Compressive Strength of Accelerated-Cured Concrete Test Specimens*. New Delhi: Bureau of Indian Standards, 2004.

[12] Krishna Rao, M. V., Rathish Kumar, P., and Azhar M. Khan., "A study on the influence of curing on the strength of a standard grade concrete mix," *FU Arch Civ Eng*, vol. 8, pp. 23–34, 2010.

[13] Mehmet Gesoglu, Barham Ali, Kasim Mermerdas, and Erhan Genyes, "Strength and transport properties of steam cured and water cured light weight aggregate concrete," *Const Build Mat*, vol. 49, pp. 417–424, 2013.

[14] Neelakantan, T. R., Ramasundaram, S., Shanmugavel, R., and Vinoth Ramesh, "Prediction of 28 day compressive strength of concrete from early strength and accelerated curing parameters," *Int J Eng Technol*, vol. 5, pp. 1197–1201, 2013.

[15] Yanbo Liu, "Effect of concrete with high volume pozzolonas-resistivity, diffusivity and compressive strength," Thesis submitted, Florida Atlantic University, Boca Raton, Florida, December 2012.

40. Shear Behavior of Concrete Deep Beams Reinforced With Steel and GFRP Bars

Ramadass S[*] and Job Thomas

Division of Civil Engineering, School of Engineering, Cochin University of Science and Technology, Kochi, Kerala, India

[*]Corresponding author: ramdas@cusat.ac.in

ABSTRACT: This paper deals with the study on the feasibility of replacing steel bars with glass fiber-reinforced polymer (GFRP) bars in concrete deep beams by comparing the experimental results of the shear strength and the mid-span deflection of six concrete deep beams longitudinally reinforced with steel rebars and the corresponding beams longitudinally reinforced with GFRP bars without stirrups. The variables considered are longitudinal reinforcement ratio and shear span-to-depth ratio . These variables are made common both for the steel and the GFRP bar-reinforced concrete test beams. The comparison between the steel and the GFRP bar-reinforced test beams on the ultimate strength, the corresponding mid-span deflection, and the load-deflection response are presented in this paper and conclusions are arrived.

KEYWORDS: Concrete Deep Beams, Steel Rebars, GFRP Rebars, Strength and Deflection, Comparison

1. INTRODUCTION

Though steel and fiber-reinforced polymer (FRP) rebars are frequently utilized in shallow reinforced concrete beams, limited research has been done on the viability of using FRP bars in reinforced concrete deep beams. This paper addresses this lacuna. Thomas and Ramadass (2019) tested six glass fiber-reinforced polymer (GFRP) bar-reinforced concrete (RC) deep beams with moderate strength concrete and collected other beam data from various literature and proposed a numerical model to predict the strength and the deflection of FRP RC deep beams. Thomas and Ramadass (2021) tested six GFRP bar RC deep beams with normal strength concrete and collected other beam data from various literature and proposed new analytical models to predict the strength and the deflection of FRP RC deep beams. Thomas and Ramadass (2021) tested 12 steel RC deep beams and collected other beam data from various literature and proposed new analytical models to predict the strength and the deflection of steel RC deep beams. This paper gives an insight into the feasibility of using GFRP rebars in place of steel rebars in concrete deep beams, by comparing the experimental results of the strength and the deflection between the steel and the GFRP RC deep beams published in the literature by these authors.

2. EXPERIMENTAL PROGRAM

A total of 12 concrete deep beams are cast and tested under four-point loading in this study. Out of the 12 beams, 6 beams are reinforced with 16 mm diameter Fe 500 steel bars with a yield strength of 527 MPa having a modulus of 200 GPa and the remaining 6 beams are reinforced with 16 mm diameter GFRP bars with an ultimate strength of 655 MPa having a modulus of 40.8 GPa. All the beam specimens are tested at the age of 28 days. The variables involved in the experimental study are longitudinal reinforcement ratio and shear span-to-depth ratio . The grade of concrete used is M70.

Chapter 40 DOI- 10.1201/9781032657271-40

The longitudinal reinforcement ratio used is 1.70% (6#16) and 1.14% (4#16). The shear span-to-depth ratio used is 0.50, 0.75, and 1.00. The cross-sectional dimensions of all 12 concrete deep beams are same. The overall depth of the beam is 500 mm, width is 170 mm, and length is 1590 mm. The effective depth is 416 mm for all the beams. All the beams are simply supported over a span of 990 mm with 300 mm overhang beyond either support. The beams are tested under four-point loading using a 1000 kN digital beam testing machine with top loading plate and bottom bearing plate having a width of 50 mm. To measure the deflection of the beam, dial gauges (least count = 0.01 mm) are installed beneath the beam, at the mid-point of its span. Table 1 shows the details of the test beams for the investigation. The designation of the beam SM6/0.50, in which, S indicates steel rebar, M indicates moderate strength concrete, 6 indicates the number of longitudinal rebars, and 0.50 indicates shear span to depth ratio . The test setup and the schematic of the cross-section for the steel and the GFRP RC test beam with six bars are shown in Figures 1 and 2, respectively. The experimental observations recorded for the steel and the GFRP test beams at the ultimate load and the corresponding mid-span deflection are presented in Table 1.

Table 1. Ultimate load and deflection of test beams.

S. No	a (mm)	a / d	No. of 16 mm dia. main bars	(ρ) (%)	16 mm dia. steel bar-reinforced concrete deep beams			16 mm dia. GFRP bar-reinforced concrete deep beams		
					Beam ID	$\rho_{u,e}$(kN)	$\Delta_{u,e}$ (mm)	Beam ID	$P_{u,e}$ (kN)	$\Delta_{u,e}$ (mm)
1	208	0.50	6	1.70	SM6/0.50	990	1.50	GM6/0.50	900	3.10
2	312	0.75	6	1.70	SM6/0.75	800	2.28	GM6/0.75	550	3.21
3	416	1.00	6	1.70	SM6/1.00	620	2.59	GM6/1.00	460	3.62
4	208	0.50	4	1.14	SM4/0.50	980	1.85	GM4/0.50	760	3.51
5	312	0.75	4	1.14	SM4/0.75	690	2.37	GM4/0.75	520	4.11
6	416	1.00	4	1.14	SM4/1.00	525	2.85	GM4/1.00	410	4.32

a = shear span; a / d = shear span-to-depth ratio; ρ = longitudinal reinforcement ratio; Pu, e = experimental ultimate load; Δu, e = experimental ultimate deflection.

2.1. Comparison of crack pattern and failure mode

The cracks observed in the steel RC test beams and the GFRP bar reinforced test beams using M70 grade concrete are given in Figure 3. In all the test beams, the failure is mobilized by the formation of diagonal crack. However, the width of the cracks in GFRP-reinforced concrete beams is found to be higher when compared to similar steel-reinforced test beams, as a result of the low modulus of GFRP bars.

2.2. Comparison of strength between steel and GFRP RC test beams

Figure 4 shows the comparison of strength between the steel and GFRP RC test beams. The strength of steel RC deep beams is found to be higher than the corresponding GFRP RC test beams. It is found that the ultimate strength of steel RC deep beams is greater than that of similar GFRP RC deep beams. Overall, it is observed that for the same reinforcement ratio, the strength of the GFRP RC deep beams is found to be between 69% and 91% of the strength produced by the steel RC deep beams.

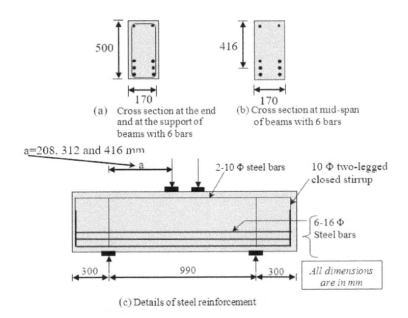

Figure 1. *Test beams with six bars, the cross-sectional diagram, details of the steel-reinforcement, and the test setup.*

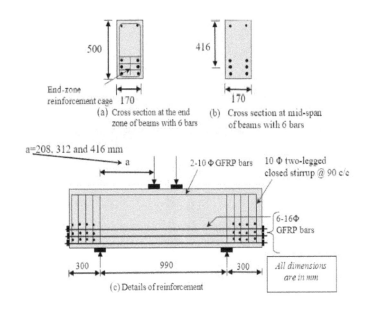

Figure 2. *Test beams with six bars, the cross-sectional diagram, details of the GFRP-reinforcement, and the test setup.*

2.3. Comparison of deflection between steel and GFRP RC test beams

Figure 5 shows the comparison of deflection between the steel and GFRP RC test beams. It is found that the mid-span deflection of the GFRP RC test beams is greater than the corresponding mid-span deflection of the steel RC test beams, at the ultimate stage. Overall, it is observed that the deflection of GFRP RC test beams is between 1.4 and 2.0 times more than the corresponding deflection of the steel RC test beams, at the ultimate stage.

2.4. Comparison of load-deflection response

Comparison of load-deflection response between the steel and the GFRP RC deep beams cast with M70 grade concrete is given in Figure 6. According to the experimental analysis, both the steel and the GFRP RC test beams exhibit a similar pattern of load-deflection response. In both steel and GFRP RC test beams, the load and the mid-span deflection are found to be almost equivalent in magnitude at the first breaking stage. When GFRP bar RC test beams are compared with equivalent steel RC test beams, it is observed that the strength and mid-span deflection magnitudes are different. At the initial stages of loading, it is found that the deformations in every test beam are steady.

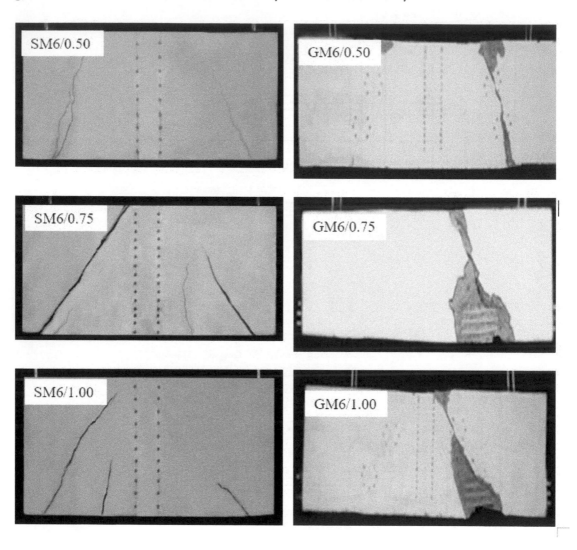

Figure 3. Crack patterns in the test beams.

However, at the later stages of the loading, it is observed that the load-deflection response of cracked beams is non-linear. This is due to the fact that the stiffness of the beams decreased as a result of the cracking. However, from Figure 6, it can be seen that the strength of steel RC test beams is greater than the corresponding GFRP RC test beams. This may be due to the localized bond slip in GFRP bar-reinforced beams, as the GFRP bars have smooth surface. The load-deflection response of the beams reinforced with steel rebars is found to exhibit higher stiffness than that of the beams reinforced with GFRP rebars. This may be due to the greater stiffness of the combined effect of the steel bars with the concrete when compared to the corresponding beams reinforced with GFRP bars.

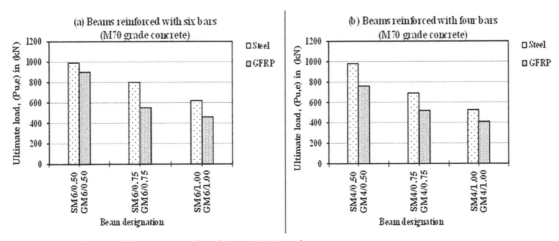

Figure 4. *Comparison of strength between steel and GFRP RC test beams.*

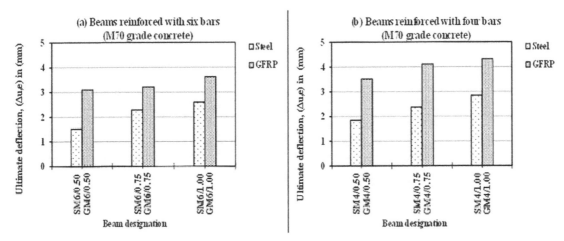

Figure 5. *Comparison of deflection between steel and GFRP RC test beams.*

Also, it can be observed that the mid-span deflection of GFRP bar-reinforced concrete deep beams is larger than the corresponding deflection of steel-reinforced concrete deep beams. It may be recalled that the average modulus of elasticity of steel rebars used in the test beams is 200 GPa and that of the GFRP rebars used is 40.8 GPa. Hence the increase in the deflection of GFRP RC deep beams may be due to the effect of the lower stiffness of the GFRP rebars with the surrounding concrete when compared to the corresponding beams reinforced with steel rebars.

3. CONCLUSIONS

The experimental study on the feasibility of substituting GFRP bars for steel bars in concrete deep beams revealed that, for the same reinforcement ratio, the strength of the GFRP RC deep beams ranged from 69% to 91% of the strength produced by the steel RC deep beams. It is found that the deflection of GFRP RC test beams is between 1.4 and 2.0 times greater than the corresponding deflection of the steel RC test beams at the ultimate stage. Hence, it may be assumed that a beam having higher GFRP reinforcement ratio can be designed to perform, as similar to that of a beam having lower steel reinforcement ratio. However, considering the fact that the GFRP bars are non-corrosive, it is expected that the total service life span of the GFRP reinforced structures would be more, when compared with that of the steel-reinforced structures. Hence, the application of GFRP bars can be recommended for the design of concrete deep beams.

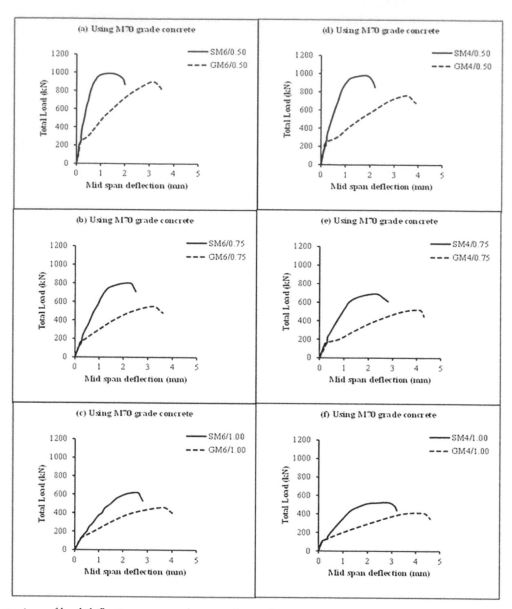

Figure 6. *Comparison of load-deflection response between the test beams.*

REFERENCES

[1] Thomas, J., and Ramadass, S., "Improved empirical model for strut efficiency factor and stiffness degradation coefficient for strength and deflection prediction of FRP RC deep beams," Struct J, vol. 29, pp. 2044–2066, 2021.

[2] Thomas, J., and Ramadass, S., "Introducing strut efficiency factor in the softened strut and tie model for the ultimate shear strength prediction of steel RC deep beams based on experimental study," Eur J Environ Civil Eng, pp. 1–39, 2021.

[3] Thomas, J., and Ramadass, S., "Prediction of the load and deflection response of concrete deep beams reinforced with FRP bars," Mech Adv Mater Struct, vol. 28, no. 1, pp. 43–66, 2019.

9 781032 656847